# Discrete Probability
## *Lecture Slide Notes*

## Ralph E. Morganstern
### *Santa Clara University*

# Table of Contents

# Table of Contents

# Table of Contents

# Table of Contents

# Preface

These Lecture Slide Notes have been used over the past several years for a two-quarter graduate level sequence in probability for engineers. Most textbooks delay presentation of some key probability concepts until later chapters where they are covered together in their discrete and continuous forms. Here we include these concepts in Part 1 on Discrete Probability so that students taking only a single quarter/semester walk away with a complete picture. Besides providing a degree of completeness for those students, this allows the transition to Part 2 on Continuous Probability to go forward with essentially no new concepts! In this manner, the increased level of mathematical sophistication encountered in the continuous domain, is not compounded by the introduction of unfamiliar concepts.

Part 1 covers counting with and without replacement, axiomatic probability models, computation techniques, conditional, joint, marginal, and Bayesian probabilities. The concept of a random variable (RV) is fully characterized by a discrete probability mass function (PMF) and a quasi-continuous cumulative distribution function (CDF). The numerical characterization of a RV is given by its mean, variance, and expectation value. Pairs of RVs give way to new concepts such as independence, covariance, and the effects of linear and bi-linear transformations. Common discrete PMFs are discussed in terms of related RV pairs, tree diagrams, and algebraic representations.

Concepts are illustrated with many examples, where the emphasis is on multiple visualizations using trees, graphical illustrations, Venn diagrams, comparison tables, and coordinate axis transformations. Part 2 covers the same concepts for continuous RVs and requires the full mechanism of differential and integral calculus as well as matrix algebra.

This "Lecture Slide Notes" format is convenient for self-study because it covers the subject matter in a concise and easily accessible manner by employing multiple visualization techniques in slide format together with focused explanatory notes. Each slide stands alone as a "one-page synopsis" that encapsulates a complete concept, algorithm, or theorem using a combination of equations, graphs, diagrams, and comparison tables. The explanatory notes are placed directly below each slide in order to reinforce and/or give additional insight into the particular technique or concept illustrated in that slide.

A Table of Contents serves to organize the slides in terms of the main probability topics covered and gives a list of all slide titles with their page numbers. An index is also provided to link related aspects of topics and cross-reference key concepts, specific applications, and the various visualization aids. Although no problem sets have been included in these notes, a good number of examples are worked out in detail. References to a number of standard text books are given, but there has been no attempt to make an exhaustive bibliography.

# 1 Paradoxes, Pitfalls, Parameters, and Parsing

## 1.1 *Preliminary Remarks*

# Preliminary Remarks

- **Probability roots** Gambling- Mere, LaPlace, Bernoulli, Gauss, Markov, ...
- **Visualization is important**
  - Trees, Venn Diagrams,
  - "boxes" or holders, Graphical sum difference coordinates
- **Easy to go astray** - Logical Arguments --> Paradox
- **Three Prisoner Paradox**:
- **Three prisoners, A,B, and C,** have been tried for murder and their verdicts will be read and their sentences executed tomorrow morning. They know that **only one of them will be declared guilty** and will be hanged to die while the other two will be set free. The **identity** of the condemned prisoner is **revealed** to the very reliable **prison guard** but not to the prisoners themselves.
- In the middle of the night, **Prisoner A** calls the guard over and makes the following request: "*Please give this letter to one of my friends - to the one who is to be released. You and I know that at least one of them will be freed.*" The guard takes the letter and promises to do as told. An hour later, **Prisoner A** calls the guard again and asks, "*Can you tell me which of my friends you gave the letter to? It could give me no clue regarding my own status because, regardless of my fate, each of my friends had an equal chance of receiving my letter.*" The guard answers, "*I gave it to* **Prisoner B***; he will be released tomorrow.*"
- *Prisoner A turns to his bed and thinks, " Before I talked to the guard, my chances of being executed were one in three. Now that he has told me that B will be released, only C and I remain, and my chances of dying have gone from 33.3% to 50%. What did I do wrong? I made certain not to ask for any information relevant to my own fate...".*
- "...Worse yet, by symmetry, my chances of dying would have risen to 50% even if the guard had named C instead of B - so my chances must have been 50% to begin with. I must be hallucinating. ..."

It is important to understand that probability arguments must be taken in context; otherwise seemingly logical arguments can lead to a paradox. The "Three Prisoner Paradox" is an example of what can go wrong if one is not careful.

Three prisoners A, B, and C are being held and one will be found guilty and executed. Thus each prisoner knows that the *a priori* probability to be found guilty is 1/3. Prisoner A tries to get information by giving a note to the guard and asking him to give it to the prisoner who will be set free. He subsequently asks the guard "to whom did you give the note" and the guard answers "I gave it to B, he will be set free". Prisoner A reasons that only he and C remain and thus his probability of being declared guilty has increased form 1/3 to 1/2 by simply asking a question that could not possibly have given him any information as to his own fate. The next slide details the subtle arguments that explain why he has reached this false conclusion.

## 1.1.1  3 Prisoner Paradox Analysis

# 3 Prisoner Paradox Analysis

- **Bayes' Theorem:** *Measurement Update*
  - Two events {0, 1}
  - *a priori* probability P(0)=P(1)=1/2
  - Two measurements {$M_0$, $M_1$}

$$P(1 \mid M_1) = \frac{P(M_1 \mid 1) \cdot P(1)}{P(M_1)}$$

$$P(M_1) = \underbrace{P(M_1 \mid 0)}_{\substack{\text{False Positive} \\ \text{Measurement} \\ \text{Statistic}}} \cdot \underbrace{P(0)}_{\substack{a\ priori}} + \underbrace{P(M_1 \mid 1)}_{\substack{\text{Positive} \\ \text{Measurement} \\ \text{Statistic}}} \cdot \underbrace{P(1)}_{\substack{a\ priori}}$$

- **Analysis# 1:** *Direct Question*
  - A= event "A is guilty"
  - $B^c$ = event "B not guilty; set free"

If A guilty, $B^c$ (not guilty) with certainty

$$P(A \mid B^c) = \frac{P(B^c \mid A) \cdot P(A)}{P(B^c)} = \frac{1 \cdot \frac{1}{3}}{\frac{2}{3}} = \frac{1}{2}$$

- **Analysis#2:** *Indirect Question*
  - A= event "A is guilty"
  - $G_B$=event guard says "B released",

If A guilty Guard can respond either B or C free with equal probability

$$P(A \mid G_B) = \frac{P(G_B \mid A) \cdot P(A)}{P(G_B)} = \frac{\frac{1}{2} \cdot \frac{1}{3}}{\frac{1}{2}} = \frac{1}{3} = P(A)$$

*If asked guard instead "will B die tomorrow", then analysis#1 is correct since the guard is now actually giving a response to a specific question whose answer contains "new information".*

**Bayes' rule** updates the probability of an event state "1" (or "A is guilty") on the basis of measurements supporting or not supporting the state. Prior to any measurement the probability is assumed to be the same for each outcome state 1/2 for binary data (1/3 for the guilt of each prisoner). Only a measurement can change this "a priori" probability. In the statement of the three prisoner paradox, the guard was *not asked a direct question* about the guilt of prisoner B and his response "I gave the note to B" does not give any new information since he could just as well have replied "I gave the note to C" with equal probability; this corresponds to Analysis#2 in which the probability that A is guilty does not change (remains 1/3). If instead, the guard was asked a *direct question* "is B guilty?" then his answer gives new information and Analysis#1 gives the correct result 1/2.

**Analysis#1:** Here the probability that "B is innocent given A is guilty" P($B^c$|A) must clearly be unity since only one prisoner can be guilty. The *a priori* probability for guilt of A is P(A) = 1/3 and that for the innocence of B is P($B^c$) = 1 - 1/3 = 2/3; substitution of these values yields the updated probability that A is found guilty P(A| $B^c$) = 1/2. New information has increased the probability!

**Analysis#2:** Since A is asking the question, the guard can only answer with a statement $G_B$ about B or $G_C$ about C and these are equally likely, so we have P($G_B$) = P($G_C$) =1/2. The guard's answer cannot depend upon the guilt or innocence of A, only on that of B and C; this means that conditioning on A is irrelevant so P($G_B$ |A) = P($G_B$) = 1/2. Substituting these values yields the updated probability for A's guilt as P(A|$G_B$) =1/3, which is the same as its *a priori* value P(A) = 1/3. As expected, this "so-called measurement" yields no new information and hence leaves P(A) unchanged. We will later learn that if the conditional probability P(A|$G_B$) equals P(A), then the two events A and $G_B$ are said to be independent, which means that one contains no information about the other.

## 1.1.2  3 Prisoner Paradox - Tree Analyses

# 3 Prisoner Paradox - Tree Analyses

**Analysis# 1:** *Direct Question – Yes  New Information; Probability changes*

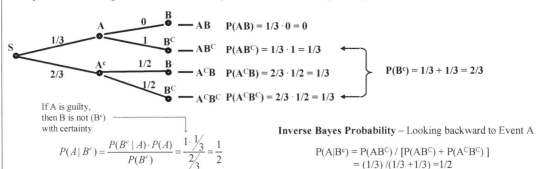

If A is guilty, then B is not ($B^c$) with certainty

$$P(A \mid B^c) = \frac{P(B^c \mid A) \cdot P(A)}{P(B^c)} = \frac{1 \cdot \frac{1}{3}}{\frac{2}{3}} = \frac{1}{2}$$

**Inverse Bayes Probability** – Looking backward to Event A

$$P(A|B^c) = P(AB^c) / [P(AB^c) + P(A^CB^c)]$$
$$= (1/3) / (1/3 + 1/3) = 1/2$$

**Analysis#2:** *Indirect Question – No  New Information ; Probability remains the same*

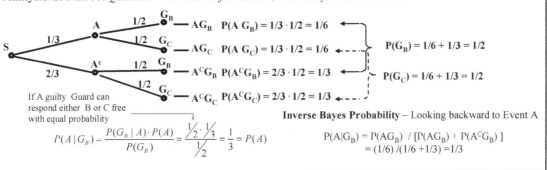

If A guilty  Guard can respond either  B or C free with equal probability

$$P(A \mid G_B) = \frac{P(G_B \mid A) \cdot P(A)}{P(G_B)} = \frac{\frac{1}{2} \cdot \frac{1}{3}}{\frac{1}{2}} = \frac{1}{3} = P(A)$$

**Inverse Bayes Probability** – Looking backward to Event A

$$P(A|G_B) = P(AG_B) / [P(AG_B) + P(A^CG_B)]$$
$$= (1/6) / (1/6 + 1/3) = 1/3$$

The two trees have identical *a priori* probabilities P(A) =1/3 (A guilty) and P($A^c$) = 2/3 (A not guilty). Their $2^{nd}$ branches give conditional probabilities for the two different sets of events {B, $B^c$} and {$G_B$, $G_C$} defined in the two analyses.

**Tree#1  Analysis:** The guard answers a direct question about the guilt or innocence of prisoner B and therefore gives specific information to prisoner A.  Note that C does not appear explicitly in the tree but belongs to the set "not B" (denoted $B^c$.)

(i) Two branches emanate from Event A and their probabilities must sum to unity. The upper branch gives the conditional probability that "B is guilty, given A is guilty" P(B|A) which must clearly be "0" since only one prisoner is guilty; the lower branch gives the conditional probability that" B is not guilty, given A is guilty" P($B^c$|A) which must be "1" for a branch sum of unity.

(ii) Similarly for the two branches emanating from the complementary event $A^c$; the upper branch gives the conditional probability that "B is guilty, given A is not guilty" P(B|$A^c$) which is "1/2" because if A is not guilty, then B is equally likely to be guilty or innocent. The lower branch gives the conditional probability that "B is not guilty given A is not guilty" as P($B^c$|$A^c$) =1/2  for a branch sum of unity. (Note that a similar analysis can be made for a direct question about prisoner C).

**Tree#2  Analysis:** The guard responds to an indirect question "who did you give the note to, *i.e.,* who will be set free?" In this case the guard's answer B or C provides no information to prisoner A and his probability remains at 1/3.

(i) Two branches emanate from Event A and their probabilities must sum to unity. The upper branch gives the conditional probability that "the guard says prisoner B will be set free, given A is guilty" P($G_B$|A) which must clearly be "1/2" since the guard has two equally likely possibilities $G_B$ or $G_C$; the lower branch gives the conditional probability "C will be set free, given that A is guilty" P($G_C$|A) which must be "1/2" for a branch sum of unity.

(ii) Similarly for the two branches emanating from Event $A^c$; the upper branch gives "the conditional probability that the guard says prisoner B, given A is not guilty" P($G_B$| $A^c$) which equals "1/2",  because if A is not guilty, then B is equally likely to be guilty or innocent.  The lower branch gives "the conditional probability that the guard says C, given that A is not guilty"  P($G_C$| $A^c$) which equals "1/2". This analysis properly takes into account the two equally probable responses the guard could have given and yields a conditional probability of guilt P(A|$G_B$) =1/3 which is the same as the *a priori* guilt P(A)=1/3 ; *i.e.,* the "measurement" yields no new information and leaves P(A) unchanged.

**Inverse Bayesian Probability** looks backwards from the outcome states of the tree in order to determine the guilt or innocence of A; for example, in Tree#1 the outcome $B^c$ occurs in two final states, {$AB^C$} and {$A^CB^C$} emanating from A and $A^C$ respectively.  Thus the "inverse" Bayesian probability P(A|$B^C$) is the probability ratio of the desired path  to the sum of paths, *viz.,*

**Tree#1**    P(A|$B^C$) = P(A$B^C$) / [P(A$B^C$) + P($A^C B^C$) ]    = (1/3) /(1/3 +1/3) =1/2

A similar argument applies to Tree#2 and yields the path probability

**Tree#2**    P(A|$G_B$) = P(A$G_B$) / [P(A$G_B$) + P($A^C G_B$) ]    = (1/6) /(1/6 +1/3) =1/3

## 1.2 Some Pitfalls

# Some Pitfalls

- **Pitfall#1:** When facts are formulated incorrectly, even correct probability analysis can lead to counter-intuitive results as in the "Three Prisoners Paradox." (Most paradoxes involve incorrect formulation of the problem)

- **Pitfall#2:** Carpenter: Measure twice and cut once.
  Probability: Calculate twice and predict once.

- **Pitfall#3:** Do not try to impress people with the statement: "There is a *finite probability* that ..."
  - By definition probability is always finite $\Pr(X) \, \varepsilon \, [0,1]$
  - What you really mean to say is " *a small non-zero probability*"

- **Pitfall#4:** Never add or subtract sets with "±" or multiply with "*". Unions "∪" and intersections "∩" are used instead. "A-B" means *A and not B*: $A \cap B^c = AB^c$

- **Pitfall#5:** Operations on sets can never yield the number "0"; if the set is *empty* you mean to write "$\phi$" (null set), not "0"!

- **Pitfall#6:** Probabilities evaluated on sets yield *numbers, not sets*; probability can be zero, *but* is *never negative*.

- **Pitfall#7:** Mutually exclusive events *are not independent*; in fact they are *totally dependent* as the existence of one excludes the other.

- **Pitfall#8:** Plural of "die" is not "dies", rather it is "dice"

The "Three Prisoner Paradox" should be ample warning that great care must be exercised when applying probability arguments to real situations. It is very easy to go astray, so solving the problem by two "independent and well reasoned" methods is a good way to safeguard against erroneous results, though it is of course not failsafe.

Also be cognizant of the difference between *sets* which are groups of elements and *numbers* which are magnitudes; the result of set operations is always another set (*not a number*). Probability is a number evaluated over sets; therefore, first manipulate the set expression into a useful form and then evaluate the probability as a number in the interval [0, 1]. If the result of a long calculation yields a negative probability, you have undoubtedly made a mistake somewhere in your calculation; check it!

Finally, most of the slide bullet statements are "head nodders" in the sense that we all know them to be true; however, we may all *fall victim to these pitfalls* if we are not mindful of the obvious.

## 1.3 *Parameters Label Distinguishable Objects*

# Parameters Label Distinguishable Objects

- **Parameters:** Blue, Ford, numbered 56, traveling at 40 mph
  - 4 distinguishing parameters: Color, Brand, Number, Speed.
  - If **do not care** about "Car Brand": Hummer, VW, Ford, ... , then collapse parameter space from 4 to just 3
  - Conversely, if we **do care** about direction **N,S,E,W**, then we expand parameter space from 4 to 5

- **Parameter Space:**
  - distinguishes objects and
  - defines how we count

- **Atomic Events:** distinguishable outcomes specified by a unique set of independent parameters or points in n-dimensional coords.

- **Compound Event:** *set theoretic definition* using "and" "or", "not", "complement": Red, VW, 44-56, N, @30-40Mph

Parameters serve as "coordinate labels" in a multi-dimensional parameter space and their range of discrete values "index" all possible outcomes. Assigning a single discrete value to each of the coordinate parameters defines a single point in parameter space called an "atomic event". On the other hand, assigning a range of discrete values to each coordinate parameter, specifies a group of points (set) that collectively define a "compound event".

If we are not interested in one of the parameters say, car brand, then summing over the car brand values collapses the parameter space to one that has one less dimension. Conversely, adding a new characteristic to the atomic event description increases the dimension of the parameter space.

## 1.4 *Parsing Phrases in Word Problem Statements*

---

### Parsing Phrases in Word Problem Statements

- **Typical Phrases**: assume X has outcomes 0,1,2,…,N
  - X is **at least** 3 means: X=3, or 4, or 5,… ,or N
  - X is **greater than** 3 means: X=4, or 5,… ,or N
  - X is **at most** 3 means: X=0, or 1, or 2, or 3
  - X is **exactly** 3 means X=3 only

- **Calculation Shortcut**
  - **Sum Prob(all outcomes) = 1:** *or* $\sum P(X) = 1$
  - Prob for X > 0 is computed directly by summing terms
    $$P(X>0) = P(X=1) + P(X=2) + P(X=3) + \cdots + P(X=N)$$
    or more simply using shortcut:   $P(X>0) = 1 - P(X=0)$

- **Probabilities of Parsed Phrases**
  - **P("*at least one*") = P(A∪B∪C)**   their union
  - **P("*not any*") = 1 – P(A∪B∪C)**   *not* at least one
  - **P("*not any*") = P (A$^c$B$^c$C$^c$)**    intersect complements
  - **P("*only A*") = P (AB$^c$C$^c$)**     intersect A, not-B, not-C

---

When reading and interpreting word problems, care must be taken to parse out phrases such as "at least", "at most", "exactly", *etc.* . Since the sum of probabilities for all outcomes must be unity, we may compute the probability for "at least three" by computing the probabilities for 0, 1, 2 and subtracting their sum from unity as a shortcut.

Moreover, we shall find that set theoretical constructs involving unions, intersections, and complements provide a natural method for parsing word statements. Given a sample space of three events {A, B, C}, we can immediately interpret the probability of "at least one" as the probability of their union. The probability of "not any" may be interpreted in two ways: (i) 1 minus probability of their union, or (ii) the probability of the intersection of their complements. The probability of "only A" is naturally written as the intersection of A with B$^c$ (not-B) and with C$^c$ (not-C). Set theory is thus a natural language for translating word statements into precise mathematical constructs on which the probability can be evaluated.

A properly drawn tree diagram can also be very useful in this respect since all outcomes are displayed and the particulars of the word statement can be picked off by inspection.

# 2 Probability and Counting

# Probability and Counting

## 2.1 *Sample and Event Spaces*

# Sample and Event Spaces

- **Probabilistic Interpretation of Random Experiments (P)**
  - *Outcomes:* sample space
  - *Events:* collection of outcomes (set theoretic)
  - *Probability Measure:* assign number "probability" $P \varepsilon [0,1]$ to event
- **Dfn#1-Sample Space (S):** Fine-grained enumeration (atomic - parameters)
  - List all possible outcomes of a random experiment
  - ME - *Mutually exclusive* - Disjoint "atomic"
  - CE - *Collectively exhaustive* - Covers all outcomes
- **Dfn#2- Event Space (E):** Coarse-grained enumeration (re-group into sets)
  - ME & CE List of **Events**

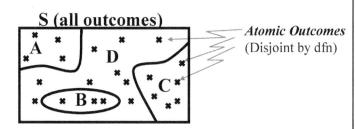

*Events: A,B,C   ME   but not CE*

*Events: A,B,C ,D   both ME   & CE*

Discrete parameters uniquely define the coordinates of the Sample Space (S) and the collection of all parameter coordinate values define all the atomic events. By their very nature, atomic events are mutually exclusive (ME) and collectively exhaustive (CE) and constitute a fundamental representation of the Sample Space S of experimental outcomes. It is often more useful to organize these atomic events into larger structures consisting of groups of related points or "events" such as the sets A, B, C, and D in the figure. Whenever we define such a set of events covering all outcomes in the sample space *without overlap* we say that they are mutually exclusive (ME) and collectively exhaustive (CE) and that they form an event space. The definition of these events is quite arbitrary and is generally set up to represent outcomes that are convenient for a particular problem.

For example, in the game of craps, there are 36 atomic events characterized by the face values $(d_1, d_2)$ of a pair of six sided dice. The events of interest are those dice combinations that have constant sums $s = d_1 + d_2$ where s labels the events with sums $\{2,3,4,5,6,7,8,9,10,11,12\}$. The sets of events labeled by their sums do not intersect as they are indexed by a unique number (ME), and moreover, they cover all possible outcomes (CE) for the pair of dice. Therefore, these sets with constant sum values constitute the Event Space for the game of craps.

## 2.1.1 Abstract Event Space Examples

# Abstract Event Space Examples

| Sample Space S | ME? | CE? | Set {A,B,C}= Event Space? |
|---|---|---|---|
| | Yes | No | No (Not CE; some samples ∉ A or B or C} |
| | Yes | Yes | Yes (Events are Disjoint) |
| | No | Yes | No (Not ME ; Events Intersect;) |

This table illustrates the concepts of mutually exclusive (ME) and collectively exhaustive (CE) for Event Spaces. The first column shows three different Sample Spaces and the two columns that follow check to see if the set consisting of events A, B, and C satisfy the "ME" and "CE" criteria for an Event Space; the last column states the test result and makes a brief comment. Inspecting the table, we see that the defined events in rows #1 and #2 do not intersect so they satisfy the "ME" criterion, while those in rows #2 and #3 cover the sample space and satisfy the "CE" criterion. (Note that in row #1 some atomic points are outside the three defined event sets A, B, C so they are not CE.) Clearly only row #2 satisfies both criteria (ME, CE) and qualifies as an event space.

Such event space decompositions are fundamental to computing probabilities because
(i)      there are no overlaps between events, so we simply add individual probability contributions,
(ii)     the whole sample space is covered, so the sum of individual event probabilities adds to unity.

This will prove to be important in future discussions, so it bears repeating: *The collectively exhaustive (CE) property assures that we cover all possible outcomes, while the mutually exclusive (ME) property guarantees that there is no double counting; thus, a direct sum of the individual probabilities over an Event Space will always be unity.*

## 2.1.2 Event Space Representations - Fair Dice

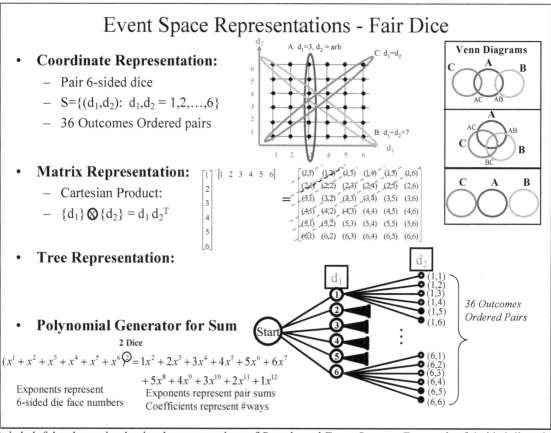

It is helpful to have simple visual representations of Sample and Event Spaces. For a pair of 6-sided dice, the **coordinate**, **matrix**, and **tree** representations displayed on this slide are all useful. **Venn diagrams** for two or three events give a visualization for generic events that is simple and obvious. In the top figure showing the $(d_1, d_2)$-coordinate representation for a pair of 6-sided dice, we use a Venn diagram overlay to visualize the events A: $\{d_1 = 3$ and $d_2 = $ arbitrary$\}$, B= $\{d_1 + d_2 = 7\}$, and C= $\{d_1 = d_2\}$. The Venn overlay makes the intersection properties of these events quite transparent: both A and B and A and C intersect, *albeit* at different points, while B and C do not intersect (no point corresponding to s = 7 and $d_1 = d_2$). Venn diagrams for more than three intersecting sets, becomes problematic as the advantage of visualization is muddled by the increasing number of overlapping regions (see next two slides).

Another interesting representation is the **polynomial generator** for the sums of a pair of 6-sided dice. The generator is the product of two 6th degree monomials with fair dice coefficients of "1", *viz.*, $(1x^1+1x^2+1x^3+1x^4+1x^5+1x^6) \cdot (1x^1+1x^2+1x^3+1x^4+1x^5+1x^6)$. Expanding this expression gives a 12[th] degree polynomial whose *exponents* represent all possible dice sums S=$\{2,3,4,5,6,7,8,9,10,11,12\}$ and whose *coefficients* C = $\{1,2,3,4,5,6,5,4,3,2,1\}$ represent the number of ways each sum can occur. The coefficients are also easily picked off the matrix representation by counting points along each dashed (red) diagonal lines representing sums. Dividing the coefficients C by 36 (total number of outcomes), yields the "probability distribution" for the pair of dice. It is interesting that the dice generator can be re-written as the product $[x(x+1)(x^2+x+1)] \cdot [x(x+1)(x^2+x+1)(x^2-x+1)^2]$ which leads to non-identical **Sicherman dice** with face values $\{1,2,2,3,3,4\}$ and $\{1,3,4,5,6,8\}$, but the **same** distribution! This is seen as follows:

$1^{st}$ term = $[(x^2+x)((x^2+x)+1)] = (x^2+x)^2 + (x^2+x) = x^4+2x^3+x^2 + (x^2+x) = 1x^1 + 2x^2 +2x^3 + 1x^4$ → $\{1,2,2,3,3,4\}$

$2^{nd}$ term = $[x(x+1)\{((x^2+1) + x)((x^2+1) - x)\}((x^2+1) - x)] = \{(x^2+1)^2-x^2\}\{(x^2+x)(x^2 - x+1)\}$

$= \{x^4+2x^2+1-x^2\}\{x^4-x^3+x^2+x^3-x^2+x\} = \{x^4+x^2+1\}\{x^4 +x\} = \{x^8+x^6+x^4+x^5+x^3+x^1\}$  → $\{1,3,4,5,6,8\}$

## 2.1.2.1 Venn Diagram for 4 Sets

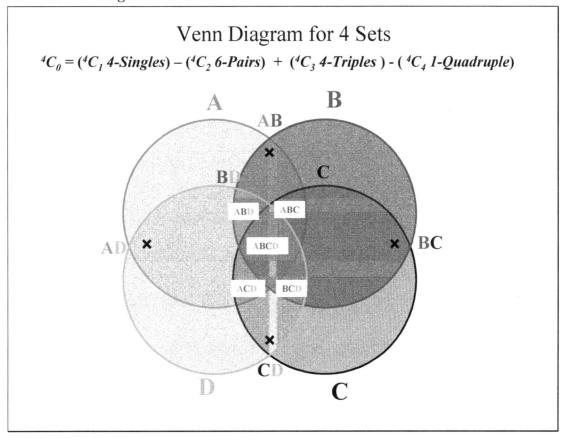

Venn Diagram for 4 Sets

$$^4C_0 = (^4C_1 \text{ 4-Singles}) - (^4C_2 \text{ 6-Pairs}) + (^4C_3 \text{ 4-Triples}) - (^4C_4 \text{ 1-Quadruple})$$

As we go to Venn diagrams with more than 3 sets the labeling of regions becomes a practical limitation to their use. In this case of 4 sets A,B,C,D, the labeling is still pretty straightforward and appears to be usable. The expression above the Venn diagram gives the number of *singles, pairs, triples, and quadruples* formed from the 4 sets (see binomial identity (3) of Slide# 2-22).

**The 4 singles** A,B,C,D are labeled in an obvious manner at the edge of each circle.

**The 6 pairs** AB, AC, AD, BC, BD, and CD are labeled at the intersection of two circles.

**The 4 triples** ABC, ABD, BCD, and ACD are labeled within "curved triangular areas" corresponding to the intersections of three circles.

**The 1 quadruple** ABCD is labeled within the unique "curved quadrilateral area" corresponding to the intersection of all four circles.

This would appear to be a useful representation for 4 events since all subsets are easily picked off the diagram; however, although the component parts of this Venn diagram are CE, the singles, pairs, and triples overlap one another and are therefore not ME. Although this presents no problem in displaying atomic points (marked with "**X**"s) for the four pairs AB, BC, CD, and AD, it falls short because we cannot "physically" display atomic points that are a member of *only* BD or a member of *only* AC. This latter fact makes the Venn diagram a less useful visualization tool. (See Slide#3-15 for a more detailed explanation.)

## 2.1.2.2 Venn Diagram for 5 Sets

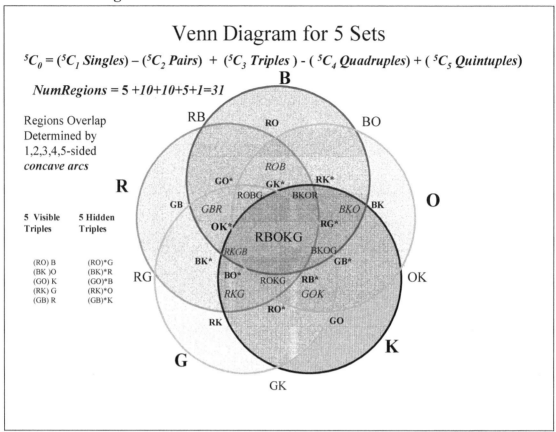

In this case of 5 sets A, B, C, D, E the labeling is quite cumbersome and appears to be unusable because half of the triples are "hidden" within other triples and quadruples. The expression above the Venn diagram gives the number of *singles, pairs, triples, quadruples, and quintuples* formed from the 5 sets (see binomial identity (3) of Slide# 2-22).

**The 5 singles** are labeled by their colors: B(lue),O(range), (blac)K, G(reen), R(ed).
**The 10 doubles** are labeled by the pairs of intersecting circles
**The 10 triples** are "curved triangular regions" formed by concave arcs of the appropriate colors, but have become a problem to label since they overlap other "tuples". The table somewhat arbitrarily divides them up into visible and hidden triples because although all are visible some are more easily discerned than others.
**The 5 quadruples** are "curved quadrilateral regions" formed by concave arcs of the appropriate colors
**The 1 quintuple** is the unique "curved pentagonal region" formed by all 5 circles appropriately in the center of the diagram.
Since the idea of a Venn diagram is to clarify things visually, placing numbers of elements in the various regions and quickly identifying them would be difficult at best. There are higher order Venn Diagrams on the Mathematica® website, but there seem to be no practical applications available. For more than three events, we shall revert to partial trees as well as non-visual algebraic methods to be discussed under inclusion/exclusion analysis (Slide#3-9; also see Slide#3-27 on Man-Hat Matching Problem.)

### 2.1.3 Two 6-sided Dice Sum and Difference Coordinates

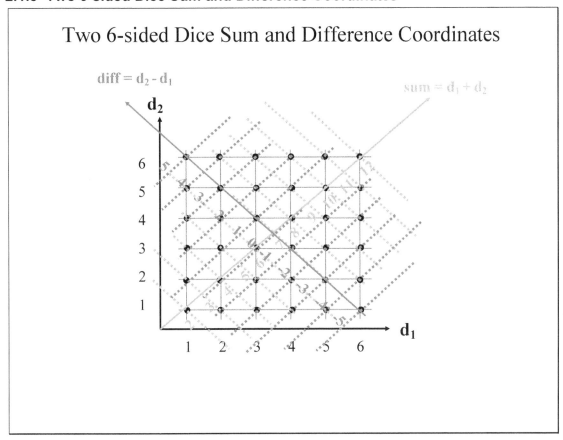

Two 6-sided Dice Sum and Difference Coordinates

In the game of craps, the sum of the two 6-sided dice determines the wins and losses according to a set of rules (Slide#2-9). Because the sum is important, the coordinate representation of the game becomes more useful if we rotate from the original $(d_1, d_2)$-coordinates to the sum and difference $(s,d)$-coordinates which are defined by the pair of equations $s = (d_1 + d_2)$ and $d = (d_1 - d_2)$. Note that the new representation is just rotated 45 from the original one.

The ***sum-axis*** (green) is rotated +45 deg from the $d_1$-axis and values along the *sum-axis* range from 2 to 12 as illustrated by the green dashed lines with -45 deg slopes. Note that lines perpendicular to the sum axis represent "surfaces" of constant sum and take on 11 values $s = \{2,3,4,5,6,7,8,9,10,11,12\}$.

The ***difference-axis*** (red) is rotated (90+45) 135 deg from the $d_1$-axis and the values along the *difference-axis* range from -5 to +5 are illustrated by the red dashed lines with +45 deg slopes. Note that lines perpendicular to the difference axis represent "surfaces" of constant difference and take on 11 values $d=\{-5,-4,-3,-2,-1,0,1,2,3,4,5\}$.

This sum-difference transformation is an example of a transformation technique that will turn out to be very useful in probability calculations involving more than one random variable and it is well worth the time spent to understand it.

## 2.1.4 Trivial Computation of Probabilities of Events

# Trivial Computation of Probabilities of Events

**Ex#1 Pair of Dice**
$S=\{(d_1,d_2):\ d_1,d_2 = 1,2,\ldots,6\}$

$E_1=\{(d_1,d_2):\ d_1+d_2 \geq 10\}$
$\quad P(E_1)=6/36=1/6$

$E_2=\{(d_1,d_2):\ d_1+d_2 = 7\}$
$\quad P(E_2)=6/36=1/6$

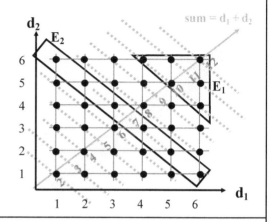

**Ex#2 Two Spins on Calibrated Wheel**
$S=\{(s_1,s_2):\ s_1,s_2 \in [0,1]\}$

$E_1=\{(s_1,s_2):\ s_1+s_2 \geq 1.5\}\ \text{-->}\ P(E_1) = \dfrac{}{1} = .5^2/2=1/8$

$E_2=\{(s_1,s_2):\ s_2 \leq .25\}\ \text{-->}\ P(E_2)=1(.25)/1=.25$

$E_3=\{(s_1,s_2):\ s_1= .85;\ s_2= .35\}\text{-->}\ P(E_3)=0/1=0$

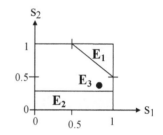

For equally likely atomic events the probability of any outcome event E is easily computed as the (#atomic events in event E)/(total # outcomes). For a pair of dice, the total # of outcomes is the product 6*6=36 and hence a simple count of the (# points in E) /36 yields the probability P(E). Two events $E_1$ and $E_2$ are defined algebraically and illustrated geometrically in the $d_1 - d_2$ plot; the probability for each event is obtained by simply counting the "dots" within each shape and then dividing by the total of 36 as shown in the slide.

The lower part of the slide shows a similar result for continuous variables. The sample space determined by two spins on a calibrated wheel [0, 1) can be represented by the continuum of all equally-probable points within the unit square in the $(s_1, s_2)$-plane. The total number of outcomes covers an area of 1*1 = 1 and hence the probability for an event E which covers "area(E)" is simply computed as the ratio of event to total area, *viz.*, P(E)= area(E) / 1.

These computations of probabilities proceeded without any formal knowledge of the subject because we had good visualizations which made the results intuitively obvious. This is not to say there is no need to study the subject, but rather to point out how simple things can appear if we have the right "picture" in mind. We shall make use of such visualization tools as is practical throughout these lecture slide notes.

## 2.1.5 Tree for the "Game of Craps"

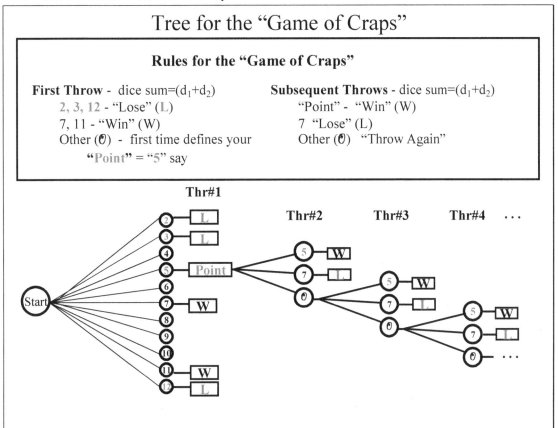

# Tree for the "Game of Craps"

### Rules for the "Game of Craps"

**First Throw** - dice sum=$(d_1+d_2)$
  2, 3, 12 - "Lose" (L)
  7, 11 - "Win" (W)
  Other (0) - first time defines your
    "Point" = "5" say

**Subsequent Throws** - dice sum=$(d_1+d_2)$
  "Point" - "Win" (W)
  7 "Lose" (L)
  Other (0) "Throw Again"

The rules of the game of craps are quite simple:

i) The first throw results in either "Win" or "Lose" (which ends the game), or establishes the "Point" and the game continues with another throw.

ii) Subsequent throws result in three outcomes Win or Lose (again stopping the game) or throw again.

The tree very nicely illustrates the game for the case in which throw#1 establishes the point as "5" and subsequent throws yield W ("5"), L ("7"), or Other ("not 5" and "not 7" ) which leads to the next throw ("throw again").

More specifically, the first throw yields one of three outcomes (i) win (throwing a "7" or "11") or (ii) lose (throwing a "2" or "3", or "12") both of which end the game or (iii) establishing the "point" (throwing one of the remaining numbers {4,5,6 ,8, 9,10}). The game continues with the next throw of the dice for which the rules are different; for the second and all subsequent throws there are again three outcomes Win (making the point) or Lose (throwing a "7") or Other (throwing any number other than the "point" or a "7").

It is seen that after establishing the "point" in the first throw, the tree continues indefinitely with the W, L, or Other branches and we must count the infinite number of nodes that result in a win W in order to compute the probability of winning with the point "5". We shall see that the win outcomes of this infinite tree results is a geometric series which can be expressed in a usable closed form expression as follows: $1+x+x^2+x^3 +... = 1/(1-x)$.

## 2.2  Set Theory Ideas

---

# Set Theory Ideas

- Formal Set Algebra
- Intersections and Unions
- DeMorgans Laws
- Translation of Words to Set Algebra

---

The idea of enumerating all possible outcomes correctly is essential to an understanding of discrete probability. If all outcomes are *equally likely* then the count of outcomes associated with a given event divided by the total number of outcomes yields a measure of the event's probability. Algebraic set theory together with visualization techniques such as trees, Venn diagrams, and coordinate graphs are useful tools to formulate these counting concepts.

The basic concepts of set algebra are most easily understood using Venn diagrams to visualize the unions, intersections, and complements used to construct compound events in the sample space. Manipulation of these constructs using commutivity, associativity, distribution of intersections over unions, and of unions over intersections, allow us to transform any word description of a desired event into a mathematical construct which can then be evaluated numerically. DeMorgan's laws for finite unions and intersections provide a unique connection between the *complement of a union* and the *intersection of the complements* that adds significantly to our ability to construct desired events.

## 2.2.1 Formal Algebra of Sets

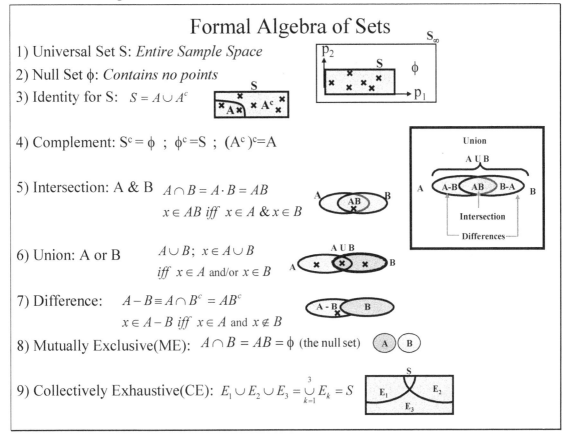

# Formal Algebra of Sets

1) Universal Set S: *Entire Sample Space*

2) Null Set $\phi$: *Contains no points*

3) Identity for S: $S = A \cup A^c$

4) Complement: $S^c = \phi$ ; $\phi^c = S$ ; $(A^c)^c = A$

5) Intersection: A & B $\quad A \cap B = A \cdot B = AB$

$\qquad x \in AB$ iff $x \in A$ & $x \in B$

6) Union: A or B $\quad A \cup B$; $x \in A \cup B$

$\qquad$ iff $x \in A$ and/or $x \in B$

7) Difference: $\quad A - B \equiv A \cap B^c = AB^c$

$\qquad x \in A - B$ iff $x \in A$ and $x \notin B$

8) Mutually Exclusive(ME): $\quad A \cap B = AB = \phi$ (the null set)

9) Collectively Exhaustive(CE): $E_1 \cup E_2 \cup E_3 = \bigcup_{k=1}^{3} E_k = S$

Set Algebra is a formal way of defining events and is very conveniently represented by Venn diagrams. The universal set S represents the entire sample space which is illustrated as the rectangle for a fixed range of the parameters along the $p_1$ and $p_2$ axes. These two parameters could be a car brand and its color or the values for pair of dice; the limited ranges along each parameter axis are fixed by the context of the problem. If there are some excluded parameter values, then they are not in the sample space and can be labeled as the null set (containing no valid points in parameter space).

The complement of a set A includes all members **not in A** and is denoted by the superscript as $A^c$. The union of any set A with its complement $A^c$ yields the whole space and this is the identity for S given in 3). Clearly the complement of the universal set S is the null set, denoted $S^c = \phi$ (since S contains all elements in the sample space its complement $S^c$ contains none).

The concepts of unions and intersections are easily visualized in terms of Venn diagrams as are the ideas of disjoint or mutually exclusive (ME) sets and covering or collectively exhaustive (CE) sets. However, for more than 3 sets a Venn diagram is nearly useless and we need to employ algebraic methods *via* a "set-theoretic" inclusion/exclusion expansion.

The difference set notation "A-B" is common, but misleading since the subtraction of two sets is of course meaningless; it makes sense only if we understand "A-B" to be "shorthand notation" for the intersection of A and not-B, which is properly written as $AB^c$. The boxed diagram summarizes these concepts.

## 2.2.2 Set Manipulation Properties

# Set Manipulation Properties

- **Intersections & Unions**

i) Commutivity: $\quad\quad\quad A \cup B = B \cup A$

ii) Associativity: $\quad\quad A \cup (B \cup C) = (A \cup B) \cup C \;\; ; \;\; A \cdot (B \cdot C) = (A \cdot B) \cdot C$

iii) Distribution of Intersection over Unions: $\quad A \cdot (B \cup C) = A \cdot B \cup A \cdot C$

Distribution of Union over Intersections: $\quad A \cup (B \cdot C) = (A \cup B) \cdot (A \cup C)$

---

- **Other Properties ( Two Sets: A , B subsets of S)**

i) Null Set: $\quad\quad\quad\quad\quad A \cdot \phi = \phi \;\; ; \;\; A \cup \phi = A$

ii) Identity Equality $\quad\quad A = A \cdot S = A \cdot (B \cup B^c) = AB \cup AB^c$

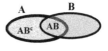

iii) Decompose into Disjoint Sets $\quad\quad A \cup B = A \cup A^c B$

$$= B \cup B^c A$$

iv) Equality of Sets $\quad A = B \;\; iff \;\; A \subset B \; \& \; B \subset A$

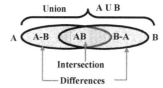

Under the two operations unions and intersections, the standard algebraic properties of sets are: commutivity, associativity, and distribution of intersection over unions and distribution of union over intersections as shown in the top panel.

The bottom panel shows a number of other important properties that illustrate the manipulation and decomposition of sets. Especially important is the idea of decomposition of two intersecting sets into a union of disjoint sets. This concept will be extremely important when we formally define probability.

## 2.2.3 DeMorgan's Laws - Finite Unions and Intersections

# DeMorgan's Laws - Finite Unions and Intersections

i) Compl(Union) = Intersec(Compls): $\quad (E_1 \cup E_2 \cup \cdots \cup E_n)^c = E_1^c \cap E_2^c \cap \cdots \cap E_n^c$

ii) Compl(Intersec) = Union(Compls): $\quad (E_1 \cap E_2 \cap \cdots \cap E_n)^c = E_1^c \cup E_2^c \cup \cdots \cup E_n^c$

**Useful Forms:**

i') Union expressed as an Intersection

$$\underbrace{(A \cup B)^c}_{\text{Compl(Union)}} = \underbrace{A^c B^c}_{\text{Intersec(Compl)}}$$

$\boxed{\text{Visualization}}$

$$\left((A \cup B)^c\right)^c = \boxed{A \cup B = (A^c B^c)^c}$$

ii') Intersection expressed as a Union

$$\underbrace{(AB)^c}_{\text{Compl(Intersec)}} = \underbrace{A^c \cup B^c}_{\text{Union(Compl)}}$$

$$\left((AB)^c\right)^c = \boxed{AB = \left(A^c \cup B^c\right)^c}$$

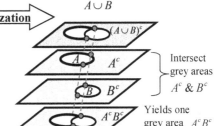

$A \cup B$

$(A \cup B)^c$

$A^c$

$B^c$

$A^c B^c$

Intersect grey areas $A^c$ & $B^c$

Yields one grey area $A^c B^c$ with A and B excluded

Taking its complement $(A^c B^c)^c$ yields white area, i.e., $A \cup B$

**DeMorgan's Laws** for the complement of finite unions and intersections states that
The complement of unions equals the intersections of the complements, and
The complement of intersections equals the union of complements
**The alternate forms** obtained by taking the complements of the original equations are often more useful because they give a direct decomposition of the union and the intersection of two or more sets. Thus we express the *union* and *intersection* directly using the altered forms:
i') The **union** equals the complement of the (intersection of complements)
ii') The **intersection** equals the complement of the (union of complements)

**A graphical construction** of $A \cup B = (A^c B^c)^c$ is also shown in the figure. $A^c$ and $B^c$ are the two shaded areas in the middle planes which exclude A and B respectively (white) ovals. Intersecting these two shaded areas excludes the white ovals, but subsequently taking the complement leaves *only the white oval areas,* which is clearly $A \cup B$.
**An intuitive proof** can be made as follows:
(i)     parse "not any" literally as "not-A and not-B" and write it as $A^c B^c$;
(ii)    parse "at least one" as "either A or B or both" and write it as $(A \cup B)$;
(iii)   write the set theory complement of $(A \cup B)$ as $(A \cup B)^c$;
But this set theory complement $(A \cup B)^c$ must also be the logical complement of "at least one", *i.e.*, it must be "not any" or $A^c B^c$, which establishes the mathematical equality $(A \cup B)^c = A^c B^c$.
A set theory proof is given on the next slide.

## 2.2.3.1 Proof: Complement(Unions) = Intersection(Complements)

# Proof: Complement(Unions) = Intersection(Complements)

- Denote the complement of the finite union as the set A

$$x \in A = \left( \bigcup_{k=1}^{n} E_k \right)^c$$

let x be an element of A;  x cannot be an element of $A^c$

Take Complement

- Since $A^c$ is just the union of all $E_k$ defn of union means that x cannot be an element of any one of the sets $E_k$

$$\Rightarrow x \notin A^c = \bigcup_{k=1}^{n} E_k$$

- Since x not a member of each $E_k$

$$\therefore \ x \notin E_k \ , \ k = 1, \cdots, n$$

it must be an element of each compl. set $E_k^c$

$$\Rightarrow x \in E_k^{\ c} \ , \ k = 1, \cdots, n$$

and therefore by definition of intersection

$$x \in \bigcap_{k=1}^{n} E_k^{\ c} \qquad\qquad QED$$

**Proof of DeMorgan's Law** (i) Complement(Unions) = Intersection(Complements) proceeds on an element-by-element basis, first identifying all elements "x" in the set A representing the *complement of unions*, and then noting that each x in A *cannot* be a member of its complement $A^c$. But $A^c$ is simply the union of the original sets, and hence if x is not a member of the union it cannot be a member of *any* individual set comprising the union; x must therefore be a member of *each* complement and hence their intersection which is the *intersection of complements*.

Out of these three forms of proof, namely, graphical or intuitive from the previous slide, or mathematical on this slide, at least one should be palatable and have some "resonance" with the reader.

## 2.2.4 Translation of Words to Set Algebra

# Translation of Words to Set Algebra

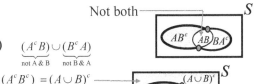

- Events A , B
- **Event $E_1$**: *Exactly one* of A or B (XOR)   $(A^c B) \cup (B^c A)$
  excludes intersection AB
  not A & B   not B & A

  $(A^c B^c) = (A \cup B)^c$
  not A & not B   (A or B) not

- **Event $E_2$**: *Neither* A or B occur
  Not both: excludes union of A& B

- **Event $E_3$**: *At least one* of A or B occur   $A \cup B$

---

- **Field of Sets** $F$ : Set of all subsets of Sample space
  Include or Exclude each *atomic element* of S   $\{s_k\}, \; k = 1, \cdots, n$

  Two choices for each yields $N = 2^n$ distinct sets   $N = (\text{total\# sets}) = 2^n$ $\begin{cases} S & \text{include all } s_k \\ \phi & \text{exclude all } s_k \end{cases}$

- **Properties of** $F$ :   Universal Set: $S \in \mathscr{F}$   $\mathscr{F} \equiv \{1 - \text{tuples}, 2 - \text{tuples}, \cdots, (n-1) - \text{tuples}, S, \phi\}$

  Complements: if $A \in \mathscr{F}$, then $A^c \in \mathscr{F}$
  n-tuple   whole   Null set
  Unions: if $A, B \in \mathscr{F}$, then $A \cup B \in \mathscr{F}$
  sets   set

- **Example of** $F$ :   $S = \{1,2,3\}, \; \mathscr{F} = \{\{1\},\{2\},\{3\},\{1,2\},\{1,3\},\{2,3\},\{1,2,3\},\{\}\}$
  $N = 2^3 = 8\, elements$

---

**Set algebra and DeMorgan's Laws** are useful tools for parsing phrases in word problem statements. For example, **"exactly one"** *or* **"only one"** of A and B is an exclusive or (**xor**) statement and can be written down literally as "(A **and** not-B) **or** (B **and** not-A)", which translates to $AB^c \cup BA^c$ meaning, "one or the other, but not both". The Venn diagram shows that it is a member of $A \cup B$ except it excludes their intersection AB.

Briefly summarizing the three events in the slide, we have
$E_1$ **"exactly one"** means "(A **and** not-B) **or** (B **and** not-A)": $AB^c \cup BA^c$
$E_2$ **"neither A or B"** means "not A **and** not B":       $A^c B^c$       $[= (A \cup B)^c$ (by DeMorgan)]
$E_3$ **"at least one of A or B"** means "A or B or both":   $A \cup B$

**Field of sets constructed from n atomic elements:** consists of all possible sets having 0, 1, 2, ..., n of the atomic elements. In forming sets from the n atomic elements we include or exclude each element in turn; thus there are a total of $N = 2^n$ such sets. This includes the null set $\phi$ containing no elements, 1-tuples, 2-tuples, 3-tuples, ...., (n-1)-tuples, and finally the n-tuple or universal set S containing all n elements.

## 2.2.5 Set Algebra Summary Graphic

# Set Algebra Summary Graphic

**Union**
$$A \cup B = A \cup A^c B$$
$$= B \cup B^c A$$

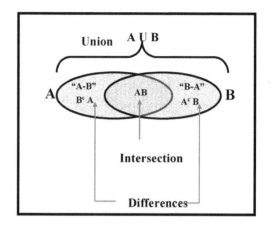

**Intersection**  $A \cap B = A \cdot B = AB$
$$x \in AB \text{ iff } x \in A \ \& \ x \in B$$

**Difference**  $A - B \equiv A \cap B^c = AB^c$
$$x \in A - B \text{ iff } x \in A \text{ and } x \notin B$$

**DeMorgans**
$$A \cup B = (A^c B^c)^c$$
$$AB = \left( A^c \cup B^c \right)^c$$

$$(A \cup B)^c = A^c B^c$$
*means*
complement of (At least one)  =  (not any)

This summary graphic illustrates algebra for two sets A and B, showing their union, their intersection, and their "difference." Again the use of the word "difference" when referring to sets must be guarded because we cannot literally take their difference; it must be understood in the context of unions, intersections, and complements!

Note that DeMorgan's Law can be interpreted as saying "*the complement of ("at least one") is "not any*." This should not be surprising since that statement was crucial to our intuitive proof of DeMorgan's law which translated parsed word statements into their mathematical equivalents according to the rules of set algebra.

Associativity and commutivity of the two basic set operations ∪ and ∩ allows extension to more than two sets using mathematical induction.

## 2.3 Basic Counting Principles

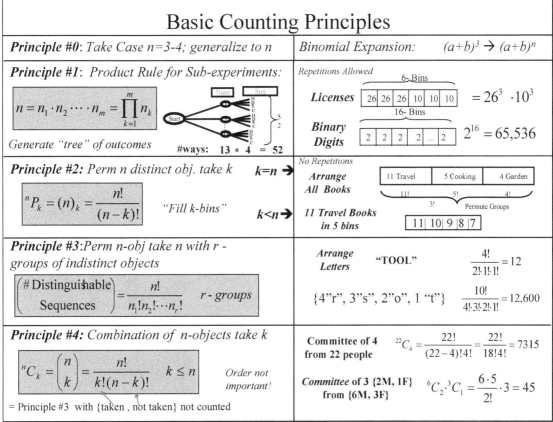

| Basic Counting Principles | |
|---|---|
| **Principle #0**: *Take Case n=3-4; generalize to n* | **Binomial Expansion:** $(a+b)^3 \rightarrow (a+b)^n$ |

**Principle #1**: *Product Rule for Sub-experiments:*

$$n = n_1 \cdot n_2 \cdots n_m = \prod_{k=1}^{m} n_k$$

*Generate "tree" of outcomes*     #ways: 13 * 4 = 52

*Repetitions Allowed*

**Licenses** 6- Bins: 26 26 26 10 10 10 $= 26^3 \cdot 10^3$

**Binary Digits** 16- Bins: 2 2 2 2 ... 2 $2^{16} = 65,536$

**Principle #2**: *Perm n distinct obj. take k*     $k=n \rightarrow$

$$^nP_k = (n)_k = \frac{n!}{(n-k)!}$$     *"Fill k-bins"*     $k<n \rightarrow$

*No Repetitions*

**Arrange All Books**: 11 Travel | 5 Cooking | 4 Garden; 11! 5! 4! / 3! Permute Groups

**11 Travel Books in 5 bins**: 11 10 9 8 7

**Principle #3**: *Perm n-obj take n with r - groups of indistinct objects*

$$\left(\begin{array}{c}\text{\# Distinguishable}\\\text{Sequences}\end{array}\right) = \frac{n!}{n_1! n_2! \cdots n_r!} \quad r\text{-groups}$$

**Arrange Letters** "TOOL"     $\frac{4!}{2! \, 1! \, 1!} = 12$

{4"r", 3"s", 2"o", 1 "t"}     $\frac{10!}{4! \, 3! \, 2! \, 1!} = 12,600$

**Principle #4**: *Combination of n-objects take k*

$$^nC_k = \binom{n}{k} = \frac{n!}{k!(n-k)!} \quad k \leq n$$     *Order not important!*

= Principle #3 with {taken, not taken} not counted

**Committee of 4 from 22 people**     $^{22}C_4 = \frac{22!}{(22-4)!4!} = \frac{22!}{18!4!} = 7315$

**Committee of 3 {2M, 1F} from {6M, 3F}**     $^6C_2 \cdot ^3C_1 = \frac{6 \cdot 5}{2!} \cdot 3 = 45$

A linear arrangement is created by selecting k objects from a set of n and placing them down in the order chosen. If the objects are distinct and their order is important, the arrangement is called a permutation; otherwise it is called a group or combination. A tree maps out all node sequences from the 1st to kth draw; *i.e.,* it enumerates all permutations of k objects drawn from n. If the order of the output sequence is unimportant, then groups are formed. Counting requires that we enumerate the number of distinct arrangements in each case. The table summarizes four basic counting principles for linear arrangements with some examples.

**Principle#0:** This is practical advice to solve a problem with n= 2,3,4 objects first and then generalize the "solution pattern" to general n and use proof by induction.

**Principle#1:** This product rule is best understood in terms of the multiplicative nature of outcomes of "independent experiments" as we "branch out" on a tree. For a single draw from a deck of cards there are 13 "number" branches and at each of these 13 number nodes we may independently choose one of four suits. The product of these two independent experiments, 13*4, yields 52 distinct cards or outcomes.

**Principle#2: Permutation (ordering) of n distinct objects** taken k at a time is best understood by setting up "k-containers" and then selecting one of "n" for the first container, one of "n-1" for the second , ... , and finally one of "n-k+1" for the kth . Thus, the total number of distinct orderings is obtained by the product rule as n*(n-1)*...*(n-k+1) = n!/(n-k)!

**Principle#3: Permutation of all "n" objects consisting of "r" groups** of indistinct objects like {3 t, 4 s, 5 u}. If all objects were distinct then the result would be n! arrangements; however, permutations within the r groups does not create a new arrangement and therefore we divide by factorials of the numbers of indistinct objects in each group to obtain n!/(n₁! n₂! ... nᵣ!).

**Principle#4: Combination of n objects** take k is related to Principles#2, #3. There are n! permutations in all, but permutations of the groups {"k-taken", "(n-k) not taken"} are unimportant so divide to find n!/(n! (n-k)!).

## 2.3.1 Permutations – Order Important

# Permutations – Order Important

- *Permutations of "n" objects from "n"*
- *Distinguishable Objects:* {x,y,z}; {#,!,3} {Book Titles various subjects}
  - Permute 3- Objects: 3 bins - any one of 3, any one of 2, 1

    | 3 | 2 | 1 |

    #outcomes = 3! => n!

  - Arrange books on shelf according to subject {11 travel , 5 cooking, 4 Garden}

    | 11 Travel | 5 Cooking | 4 Garden |

    11!    5!    4!

    3!    Permute Groups

    Permute Distinguishable Titles within Groups

  - Tree for subject group arrangements
    on shelf {x= travel , y=cooking, z=Garden }

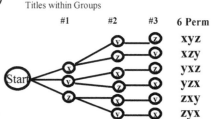

    #1    #2    #3    6 Perm
    xyz
    xzy
    yxz
    yzx
    zxy
    zyx

- *Permutations of "k" objects from "n":*

  | n | n-1 | n-2 | n-3 | ... | n-(k-1) |

  $k$ - Bins:   #1  #2   #3  #4   ...   #k

  $$^nP_k = (n)_k = n \cdot (n-1) \cdots (n-(k-1)) \cdot \underbrace{\frac{(n-k) \cdot (n-k-1) \cdots 1}{(n-k) \cdot (n-k-1) \cdots 1}}_{=1} = \frac{n!}{(n-k)!}$$

  *Principle #2: k-perm of n-obj*

  $$^nP_k = (n)_k = \frac{n!}{(n-k)!}$$

**Permutations deal with the ordering of distinct objects** (book titles). The method of ordering is key to how this gets applied. Here are some variations.

1) If we have **20 distinct book titles** then the number or orderings is simply $20! = 2.4 \cdot 10^{18}$

2) If the **20 books fall into three book categories** that must remain together, say 11 Travel, 5 Cooking, and 4 Garden we have instead $11! * 5! * 4!$ permutations of the books within their own categories and 3! orderings of the categories themselves to give $3! * (11! * 5! * 4!) = 6.9 \cdot 10^{11}$

3) **If we randomly select three groups as follows:** select 11 from 20 to place in Group#1, select 5 from remaining 9 to place in Group#2, and place the remaining 4 into Group#3. The number of orderings in this case is found by taking the product of the individual group permutations to give $^{20}P_{11} * {}^9P_5 * {}^4P_4 = (20!/9!) * (9!/4!) * (4!/0!) = 20!$, which is the same result as case 1). Note that since the groups were randomly chosen in the first place, we **do not** multiply by 3! to permute them as we did in case 2). Thus creating *random groups* is the *same as having no groups at all!*

4) **If we remove titles** and simply label all 11 Travel books with the letter "T", all 5 Cooking books with "C", and all 4 Garden books with "G", then the total number of orderings is $20!/(11! * 5! * 4!) = 21,162,960 = 2.1 \cdot 10^7$ (Principle#3)

5) **If we keep all T s , C s , and G s together in their own category,** we simply have $3! = 6$ arrangements; *i.e.*, we just permute the 3 categories since permutations of *identical objects* within a category does not constitute a new arrangement.

## 2.3.1.1 Permutations with Sub-Groups of Identical Objects

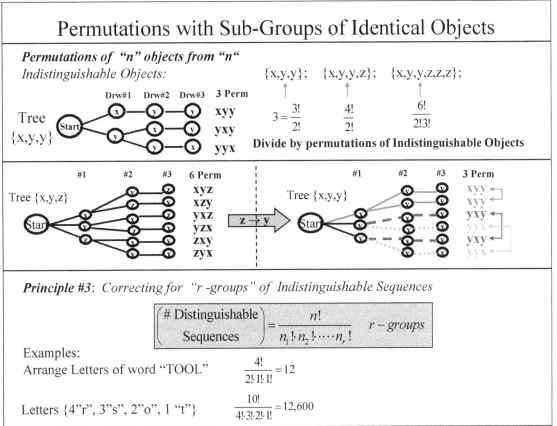

The previous slide displayed a tree of distinct objects {x,y,z} which resulted in 3!=9 orderings; however, if two objects in the original set are identical, say y=z, then {x,y,y} has 3!/2!=3 which can be verified by drawing the tree.

The **top panel** shows the three permutations {xyy, yxy, yyx} resulting from the tree {x,y,y} having one "x" and 2 "y"s; note that if order is not important, then the three distinct permutation outputs map into a single *group* consisting of 2 "y"s and one "x" (in any order).

The **middle panel** shows how the tree of distinct objects {x,y,z} maps into the tree {x,y,y} by setting z=y; the original 6 paths yield only 3 distinct permutation outputs. The pairs of identical paths shown may be combined into a single path to yield the tree in the upper panel with path sequences {xyy, yxy, yyx}.

The **bottom panel** gives some examples of Principle #3. Recall the last example in previous slide (2-18), in which the *specific book titles* were replaced the symbols "T", "C", and "G". This created three groups of identical objects, denoted by the set {11 "T", 5 "C", 4 "G"}; this set is analogous to the set created by the word "TOOL" in this slide, namely, {1 "T", 2 "O", 1 "L"}. They are both applications of Principle#3 which gives the number of distinct arrangements as the factorial of all n-objects divided by the factorials of the individual groups of indentical objects; thus, we have the following two calculations for the number of distinct arrangements:

$$\{11 \text{ "T"}, 5 \text{ "C"}, 4 \text{ "G"}\} \rightarrow (11+5+4)! / (11! \, 5! \, 4!) = 20! / (11! \, 5! \, 4!)$$

$$\{1 \text{ "T"}, 2 \text{ "O"}, 1 \text{ "L"}\} \rightarrow (1+2+1)! / (1! \, 2! \, 1!) = 4!/2!$$

## 2.3.2 Combinations - Order Not Important

# Combinations - Order Not Important

***Combinations of "k" objects from "n"***     *Distinguishable Objects:* {A,B,C}

**3 obj take 3** yields 3! = 6 Permutations of three letters:{ABC, ACB, BAC, BCA, CAB, CBA}
only single "grouping" consisting of the letters A,B,C in any order!    **$^3C_3$ =1**
Thus if order not important we divide perms by 3! leading to "one combination" group {ABC}

**3 obj take 2** yields $^3P_2 = (3)_2 = 3!/(3\text{-}2)! = 6$ Perms of two letters {*AB*, <u>AC</u>, *BA*, **BC**, <u>CA</u>, **CB**}
But only three groups as indicated by the italic (red), bold (green), and underlined (grey)
groupings **$^3C_2$ =3**
Thus if order not important we divide perms by 2!    $^3C_2 = \binom{3}{2} = \dfrac{^3P_2}{2!} = \dfrac{(3)_2}{2!} = \dfrac{3!/(3-2)!}{2!} = \dfrac{3!}{(3-2)!\,2!}$
Combination of 3 objects taken 2 at-a-time is

***Principle #4:*** *Combination of n-objects taken k at-a-time*  $^nC_k = \binom{n}{k} = \dfrac{n!}{k!(n-k)!} = \binom{n}{n-k}\quad k \le n$

***Combinations Equivalent to Permutations with 2 Groups of***
***Indistinguishable Objects: "taken" and "not-taken"***

*Split objects into two groups:*    $^nP_k = \dfrac{\text{Perm } n \text{ obj}}{\text{Perm } n\text{-}k \text{ not taken}} = \dfrac{n!}{(n-k)!}$  *Order within "not taken" group unimportant ... so divide by (n-k)!*

*n- objects*

$^nC_k = \dfrac{\text{Perm } n \text{ obj}}{(\text{Perm } n\text{-}k \text{ not taken}) \cdot (\text{Perm } k \text{ taken})} = \dfrac{n!}{(n-k)!\,k!}$

*k "taken"*    *(n-k) "not taken"*    *Order within both groups unimportant ... so divide by k! and (n-k)!*

Principle#4 describes *combinations* as the number of arrangements of k objects taken from n, where the order of the objects taken is not important. The simple examples on this slide show that the number of arrangements in this case can be expressed as the number of permutations of n objects taken k at a time, given that the order of the "k-taken" is unimportant. This expresses the combinations $^nC_k$ as permutations $^nP_k$ divided by k! or $^nC_k = {}^nP_k/k! = [n!/(n\text{-}k)!]/k! = n!/[k!\,(n\text{-}k)!]$.

Two simple examples on the slide illustrate this relationship between permutations and combinations
**3 objects take 3** yields: 3!= 6 permutations, but only 1 combination
**3 objects take 2** yields: 3!/1!= 6 permutations, but only 3 combinations (AB equivalent to BA)

We also graphically illustrate the general idea that combinations are equivalent to permutations with two groups of indistinguishable objects, namely those "taken" and those "not taken". Principle#3 immediately gives combinations as n! permutations divided by the factorials of these two groups, *viz.*,

**Combinations:** $^nC_k = n!/[(n\text{-}k)!\,k!]$ - where the order of both "taken" and "not taken" are unimportant

Permutations, on the other hand, only have one group in which order is not important, namely "not-taken", so it is determined as n! permutations divided by the factorial of this one group with n-k objects, *viz.*,
**Permutations:** $^nP_k = n!/(n\text{-}k)!$ - where only the order of "not taken" is unimportant

### 2.3.3 Binomial Theorem – Identities and Examples

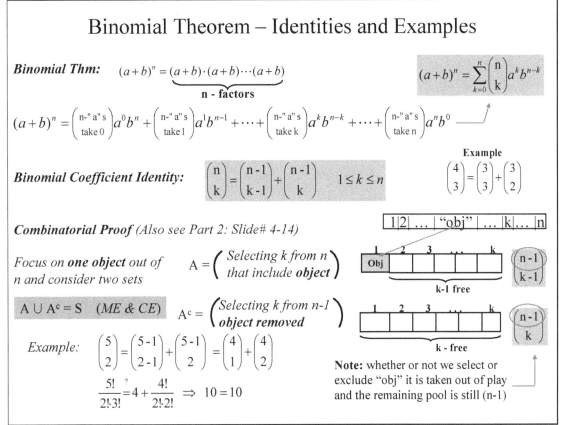

# Binomial Theorem – Identities and Examples

**Binomial Thm:** $(a+b)^n = \underbrace{(a+b)\cdot(a+b)\cdots(a+b)}_{\textbf{n - factors}}$

$$(a+b)^n = \sum_{k=0}^{n}\binom{n}{k}a^k b^{n-k}$$

$$(a+b)^n = \binom{\text{n-"a"s}}{\text{take }0}a^0 b^n + \binom{\text{n-"a"s}}{\text{take }1}a^1 b^{n-1} + \cdots + \binom{\text{n-"a"s}}{\text{take }k}a^k b^{n-k} + \cdots + \binom{\text{n-"a"s}}{\text{take }n}a^n b^0$$

**Binomial Coefficient Identity:** $\binom{n}{k} = \binom{n-1}{k-1} + \binom{n-1}{k} \qquad 1 \le k \le n$

**Example**

$$\binom{4}{3} = \binom{3}{3} + \binom{3}{2}$$

**Combinatorial Proof** *(Also see Part 2: Slide# 4-14)*

$\boxed{1|2|\ldots|\text{"obj"}|\ldots|k|\ldots|n}$

*Focus on* **one object** *out of n and consider two sets* $\quad A = \left(\begin{array}{c}\text{Selecting } k \text{ from } n \\ \text{that include } \textbf{object}\end{array}\right)$

$A \cup A^c = S \quad (ME \,\&\, CE)$

$A^c = \left(\begin{array}{c}\text{Selecting } k \text{ from } n-1 \\ \textbf{object } \text{removed}\end{array}\right)$

*Example:* $\quad \binom{5}{2} = \binom{5-1}{2-1} + \binom{5-1}{2} = \binom{4}{1} + \binom{4}{2}$

$$\frac{5!}{2!\,3!} \overset{?}{=} 4 + \frac{4!}{2!\,2!} \implies 10 = 10$$

**Note:** whether or not we select or exclude "obj" it is taken out of play and the remaining pool is still (n-1)

---

The binomial expansion of $(a + b)^n$ is the product of n factors of the term $(a + b)$. As such, it can be thought of as selecting k "a"s and (n-k) "b"s from the n product factors $(a + b)$ and then summing over the values k = 0, 1, 2, ..., n . Thus the 1st coefficient is the #ways to choose k = 0 "a"s from n which yields the term ${}^nC_0\, a^0 b^{n-0}$; similarly, the 2nd coefficient is the #ways to choose k = 1 "a" from n and yields the term ${}^nC_1\, a^1 b^{n-1}$, *etc.* . The general term, ${}^nC_k\, a^k b^{n-k}$, is summed over k = 0,1, 2, ..., n to yield the binomial expansion of $(a + b)^n$ .

There are a number of binomial identities shown on the next slide that can be proved algebraically; an alternate method of proof is a "combinatorial argument" using set theory. Here we give such a proof for one binomial identity that decomposes ${}^nC_k$ into the sum of two terms ${}^{n-1}C_{k-1}$ and ${}^{n-1}C_k$. This proof hinges on the fact that any event space S may be trivially decomposed into the union of a set A and its complement $A^c$. Hence, counting the number of ways for each mutually exclusive component and adding must give the same result as the direct computation of ${}^nC_k$ ("n take k").

If we concentrate on the selection of the "a" s from the n factors, and further pick one specific "a", say from the factor labeled "obj" in the illustration, then the set A containing this "obj" needs k-1 more "a"s to be chosen from the remaining n-1 factors; this may be done in ${}^{n-1}C_{k-1}$ ways. The complement $A^c$ does not contain the "obj" and hence that "obj" is removed from play leaving n-1 factors from which to make k choices, i.e., ${}^{n-1}C_k$. Since these two sets A and $A^c$ are ME and CE the sum of their outcomes must equal the number of outcomes from the original sample space S, thus yielding the binomial identity ${}^nC_k = {}^{n-1}C_k + {}^{n-1}C_{k-1}$ .

## 2.3.3.1 Other Binomial Identities

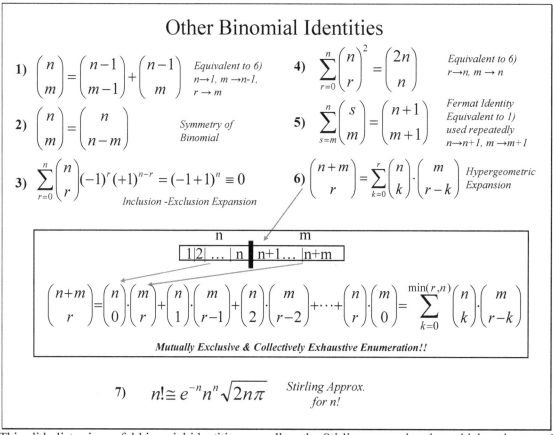

### Other Binomial Identities

1) $\dbinom{n}{m} = \dbinom{n-1}{m-1} + \dbinom{n-1}{m}$  *Equivalent to 6)*  $n \to 1,\ m \to n\text{-}1,\ r \to m$

2) $\dbinom{n}{m} = \dbinom{n}{n-m}$  *Symmetry of Binomial*

3) $\sum\limits_{r=0}^{n} \dbinom{n}{r}(-1)^r (+1)^{n-r} = (-1+1)^n \equiv 0$  *Inclusion -Exclusion Expansion*

4) $\sum\limits_{r=0}^{n} \dbinom{n}{r}^2 = \dbinom{2n}{n}$  *Equivalent to 6)*  $r \to n,\ m \to n$

5) $\sum\limits_{s=m}^{n} \dbinom{s}{m} = \dbinom{n+1}{m+1}$  *Fermat Identity Equivalent to 1) used repeatedly*  $n \to n+1,\ m \to m+1$

6) $\dbinom{n+m}{r} = \sum\limits_{k=0}^{r} \dbinom{n}{k} \cdot \dbinom{m}{r-k}$  *Hypergeometric Expansion*

$$\underset{\text{n}}{\underbrace{\boxed{1|2|\ \ldots\ |\text{n}}}}\ \underset{\text{m}}{\underbrace{\boxed{\text{n+1}\ldots\ |\text{n+m}}}}$$

$$\dbinom{n+m}{r} = \dbinom{n}{0}\cdot\dbinom{m}{r} + \dbinom{n}{1}\cdot\dbinom{m}{r-1} + \dbinom{n}{2}\cdot\dbinom{m}{r-2} + \cdots + \dbinom{n}{r}\cdot\dbinom{m}{0} = \sum\limits_{k=0}^{\min(r,n)} \dbinom{n}{k}\cdot\dbinom{m}{r-k}$$

***Mutually Exclusive & Collectively Exhaustive Enumeration!!***

7) $\quad n! \cong e^{-n} n^n \sqrt{2n\pi}$  *Stirling Approx. for n!*

This slide lists six useful binomial identities as well as the Stirling approximation which estimates n! for large values of n; your calculator uses this formula to prevent the overflow that would occur in the direct product computation of large factorials (*e.g.*, 100!).

We have given a ***proof of 1)*** on the last slide, 2) is obvious from the definition of combinations, and 3) is also obvious because the sum is precisely the binomial expansion of $(-1+1)^n$ which of course is 0.

The combinatorial ***proof of 6)*** is outlined in the boxed expansion at the bottom of the slide and ***proof of 4)*** is obtained as a special case (m=n) of 6). The ***proof of 6)*** recognizes that the combinatorial term $^{n+m}C_r$ on the LHS can be enumerated by considering m and n to be distinct pools from which objects are selected, and then enumerating all the possible ways (CE) that the "selections" can be apportioned to each. This enumeration is mutually exclusive because the events {"3 from n" and "r-3 from m"} and {"2 from n" and "r-2 from m"} do not intersect; in fact all events in the expansion are non-intersecting. Thus the sum of ME and CE terms on the RHS must be equal $^{n+m}C_r$ proving the identity.

***Proof of 1)*** follows by letting n->1, m->n-1 and r->m in 6) which yields

$$^nC_m = {}^{1+(n-1)}C_m = \Sigma^1_{k=0}\, {}^1C_k\, {}^{(n-1)}C_{m-k} = {}^1C_0\, {}^{(n-1)}C_{m-0} + {}^1C_1\, {}^{(n-1)}C_{m-1} = {}^{(n-1)}C_m + {}^{(n-1)}C_{m-1}$$

It is a good exercise to take the time to understand all these identities and their inter-related proofs.

## 2.3.3.2 Choosing Committees with and without Restrictions

# Choosing Committees with and without Restrictions

- **Committee of 4 from 22 people**    $^{22}C_4 = \begin{pmatrix} 22 \\ 4 \end{pmatrix} = \dfrac{22!}{(22-4)!4!} = \dfrac{22!}{18!4!} = 7315$    *No restrictions*

- **Committee of 3 {2M, 1F} from {6M, 3F}**    $^{6}C_2 \cdot {}^{3}C_1 = \dfrac{6 \cdot 5}{2!} \cdot 3 = 45$    *With restrictions*

- **Committee of 5 from {4M, 6F} with condition: Clyde (C) and Jake (J) not on same committee unless at least "1F" (could be "2F" or "3F")**

- **Condition implies must draw from {C, J, M1, M2}, {F1, F2, F3, F4, F5, F6} => ME & CE Events:**

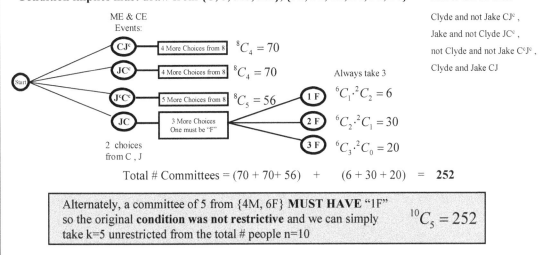

ME & CE Events:

Clyde and not Jake $CJ^c$,
Jake and not Clyde $JC^c$,
not Clyde and not Jake $C^cJ^c$,
Clyde and Jake $CJ$

$CJ^c$ — 4 More Choices from 8    $^{8}C_4 = 70$

$JC^c$ — 4 More Choices from 8    $^{8}C_4 = 70$

$J^cC^c$ — 5 More Choices from 8    $^{8}C_5 = 56$

$JC$ — 3 More Choices One must be "F"

Always take 3

1 F    $^{6}C_1 \cdot {}^{2}C_2 = 6$

2 F    $^{6}C_2 \cdot {}^{2}C_1 = 30$

3 F    $^{6}C_3 \cdot {}^{2}C_0 = 20$

2 choices from C, J

Total # Committees = (70 + 70 + 56)  +  (6 + 30 + 20)  =  **252**

Alternately, a committee of 5 from {4M, 6F} **MUST HAVE "1F"** so the original **condition was not restrictive** and we can simply take k=5 unrestricted from the total # people n=10    $^{10}C_5 = 252$

Here are some sample problems using combinations. In simple cases without restrictions, the answer may be written down directly in terms of combinations and products of combinations. In cases for which the restrictions are involved, it is wise to decompose the problem statements using set algebra to define ME and CE events and then map it out using these events in a tree as shown. The complex word statement of the problem will become more transparent as you visualize the problem using the set algebra and trees.

There is usually more than one way to do a problem and doing the problem several different ways gives you more insight into the particular problem and the process of problem solving in general. The boxed solution shows a simple way to solve the same problem by just stepping back and realizing that the original restriction of having at least 1 female on the committee was actually not a restriction at all. This is because a committee of 5 chosen from the original set of {4 M, 6 F} *must necessarily have at least* 1 F since you can only choose a maximum of 4 M. Thus the solution is just the unrestricted "from 10 take 5"

## 2.3.3.3 5 Card Draw Poker

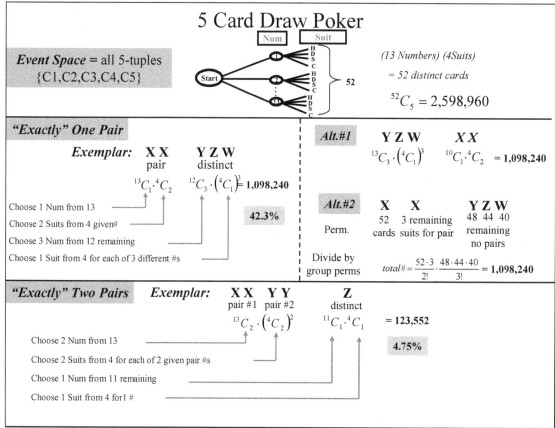

The game of five-card draw poker uses an ordinary deck of cards with 13 numbers assigned to the cards as follows: {1(Ace),2,3,4,5,6,7,8,9,10, 11(Jack),12(Queen), 13 (King)}; there are 4 suits, namely, Hearts, Spades, Clubs, and Diamonds designated by {H,S,C,D}. In one form of poker 5 cards are dealt to a hand, so the sample space is the set of all 5-tuples. Since the hand is the same no matter what order the 5 cards are arranged, the total #outcomes is the #combinations 52 take 5: $^{52}C_5$=2,598,960. Thus the probability of any hand is just the #ways to produce that hand divided by 2,598,960. A useful counting aid is to write down an "exemplar" such as this "XX YZW" pattern for a single pair with XX denoting the *same number from different suits* and YZW denoting *distinct numbers from any suit*.

**Exactly one pair:** The "exemplar" XX YZW is a simple way to characterize 1 pair.

| | |
|---|---|
| Choose 1 number from 13 | $^{13}C_1$ |
| Number is fixed so choose 2 suits from 4{H,S,C,D} | $^4C_2$ |
| Now choose 3 different #s (no more pairs) from remaining 12 | $^{12}C_3$ |
| Finally for each number choose 1 suit from 4 $\quad ^4C_1{}^4C_1{}^4C_1 =$ | $(^4C_1)^3$ |

Multiplying above four terms and dividing by $^{52}C_5$ yields **42.3% for 1 pair**

**Exactly two pairs:** The "exemplar" XX YY W characterizes 2 pairs

| | |
|---|---|
| Choose 2 numbers from 13 | $^{13}C_2$ |
| Two numbers are fixed so choose 2 suits from 4{H,S,C,D} for each | $(^4C_2)^2$ |
| Now choose 1 number from remaining 11 (for 5th card) | $^{11}C_1$ |
| Finally choose 1 suit from 4 | $^4C_1$ |

Multiplying above four terms and dividing by $^{52}C_5$ yields **4.75% for 2 pair.**

## 2.3.3.4  5 Card Draw Poker - Continued

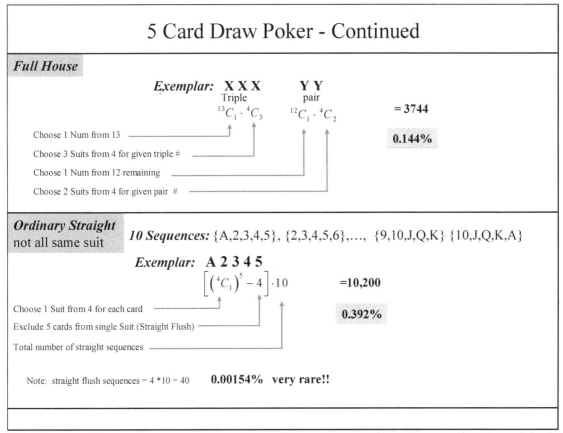

# 5 Card Draw Poker - Continued

**Full House**

*Exemplar:* **X X X**   **Y Y**
  Triple      pair
$^{13}C_1 \cdot {}^4C_3$   $^{12}C_1 \cdot {}^4C_2$   **= 3744**

**0.144%**

Choose 1 Num from 13

Choose 3 Suits from 4 for given triple #

Choose 1 Num from 12 remaining

Choose 2 Suits from 4 for given pair #

**Ordinary Straight**
not all same suit

*10 Sequences:* {A,2,3,4,5}, {2,3,4,5,6},..., {9,10,J,Q,K} {10,J,Q,K,A}

*Exemplar:*  **A 2 3 4 5**

$$\left[\left({}^4C_1\right)^5 - 4\right] \cdot 10 \qquad =10,200$$

**0.392%**

Choose 1 Suit from 4 for each card

Exclude 5 cards from single Suit (Straight Flush)

Total number of straight sequences

Note:  straight flush sequences = 4 *10 = 40     **0.00154%  very rare!!**

The full house  follows the same pattern as 1 and 2 pair; however the ordinary straight (not all in the same suit) requires a little more analysis.

**Ordinary Straight:** An ace A can be considered as either a "1" or a "13"and there are 10 possible straight sequences starting with the ace as "1" {A,2,3,4,5} and ending with the ace as "13" {10,J,Q,K,A}. Since all these sequences have the same probability we only need consider the exemplar for the first sequence {A,2,3,4,5} and multiply the result by 10.

If we make no restriction on the suits of the 5 cards, then we simply choose one suit for each number yielding $({}^4C_1)^5$ = 1024 ways. We must subtract from this the number of ways to pick all 5 cards from the same suit since this is not an "ordinary straight". The first card A can be chosen from 4 suits and the remaining cards are uniquely chosen since both their number and suit are now fixed. Thus, we subtract 4 from the previous result to obtain ( $^4C_1)^5$ - 4 = 1020 and then multiply by the 10 possible straight sequences to yield 10200 and finally the probability **0.392% for an ordinary straight.**

A **royal flush** (straight with all 5 from same suit) is a rare occurrence since there are only 4*10 = 40 ways which yields the small probability .00154%.

## 2.3.4 Binomial/Trinomial/Multinomial - Groupings

# Binomial/Trinomial/Multinomial - Groupings

**Binomial Expansion:**

$$(a+b)^n = \begin{pmatrix} n\text{-"factors"s} \\ 0\text{-"a"s, } n\text{-"b"s} \end{pmatrix} a^0 b^n + \cdots + \begin{pmatrix} n\text{-"factors"s} \\ k\text{-"a"s, } (n-k)\text{-"b"s} \end{pmatrix} a^k b^{n-k} + \cdots + \begin{pmatrix} n\text{-"factors"s} \\ n\text{-"a"s, } 0\text{-"b"s} \end{pmatrix} a^n b^0$$

*Symmetric Notation*

$$(a+b)^n = \sum_{k=0}^{n} \begin{pmatrix} n \\ k, n\text{-}k \end{pmatrix} a^k b^{n-k} = \sum_{\substack{j,k=0 \\ j+k=n}}^{n} C_{j,k}^n a^j b^k \;\; ; \quad C_{j,k}^n = \frac{n!}{j!\,k!} = \begin{pmatrix} n \\ j \end{pmatrix} = \begin{pmatrix} n \\ k \end{pmatrix}$$

*#terms in sum* $= {}^2\mathcal{C}_n = {}^{2+n-1}C_n = {}^{n+1}C_1 = n+1$

**Trinomial Expansion:**

$$(a+b+c)^n = \sum_{\substack{j,k,l=0 \\ j+k+l=n}}^{n} C_{j,k,l}^n a^j b^k c^l \;\; ; \qquad C_{j,k,l}^n = \frac{n!}{j!\,k!\,l!}$$

"$n$" distinct objects fall into "3" groups with sizes $j, k, l$

*#terms in sum* $= {}^3\mathcal{C}_n = {}^{3+n-1}C_n = {}^{n+2}C_2 = \frac{(n+2)\cdot(n+1)}{2}$

**Multinomial Expansion:**

$$(a_1 + a_2 + \cdots + a_r)^n = \sum_{\substack{n_1,n_2,\cdots n_r=0 \\ n_1+n_2+\cdots+n_r=n}} C_{n_1,n_2,\cdots n_r}^n a_1^{n_1} a_2^{n_2} \cdots a_r^{n_r} \;\; ; \quad C_{n_1,n_2,\cdots n_r}^n = \frac{n!}{n_1!\,n_2!\cdots n_r!}$$

$$C_{n_1,n_2,\cdots n_r}^n = \frac{n!}{n_1!\,n_2!\cdots n_r!}$$

"$n$" distinct objects fall into "r" groups with sizes $n_1, n_2, ..., n_r$

*#terms in sum* $= {}^r\mathcal{C}_n = {}^{r+n-1}C_n = {}^{r+n-1}C_{r-1}$

The **binomial expansion** can written using a new notation as ${}^nC_{k,(n-k)}$ = n! / [k! (n-k)!], a form that explicitly specifies the *two factorials in the denominator* and thus makes obvious the equivalence of the statements "n take k" (${}^nC_k$) and "n take (n-k)" (${}^nC_{n-k}$). Defining a new index j = (n - k), we write ${}^nC_{j,k}$ = n!/(j!k!), which is clearly symmetric in the two indices j and k (${}^nC_{k,j}$ = ${}^nC_{j,k}$). In this notation, the sum over binomial terms allows indices j and k to both range from 0 to n, but subject to the restriction j + k = n as shown explicitly in the upper panel of the slide.

This form of the binomial expansion allows an immediate generalization to a **trinomial expansion** of $(a+b+c)^n$ in terms of the trinomial coefficients defined by ${}^nC_{j,k,l}$ = n!/(j!k!l!). The sum over trinomial terms is now restricted so that the sum of indices j+k+l=n. as j,k,l each range from 0 to n.

The **multinomial expansion** generalizes the above ideas to factors having r-terms and represents the expansion of $(a_1+ a_2+ a_3+ ... +a_r)^n$ with coefficients and sum restrictions defined in the bottom panel.

These binomial, trinomial, and multinomial expansion coefficients are often encountered when distributing n-objects into groups where order is not important. These groups represent the occupancy of the two terms of the binomial $(a+b)^n$, or the three terms of the trinomial $(a+b+c)^n$, or, in general, the r-terms of the multinomial $(a_1+a_2+a_3+... +a_r)^n$. The explicit expansion of these expressions enumerate all the distinct groups at $n^{th}$ node of a tree with 2, 3, or r branches at each node. If the terms of the multinomial sum to unity $a_1+a_2+a_3+... +a_r =1$, then the resulting multinomial expansion gives the probability distribution for a whole class of events described by that multinomial and the individual $a_i \cdot$s represent the probability for each of the terms.

## 2.3.4.1 Understanding Multinomials - Examples

# Understanding Multinomials - Examples

**Identity yields Multinomial Coeff Expansion in terms of Binomial Coeffs**

$$C^n_{n_1,n_2,n_3} = \frac{n!}{n_1!\,n_2!\,n_3!} \cdot \underbrace{\frac{(n-n_1)!}{(n-n_1)!}}_{=1} \cdot \underbrace{\frac{(n-n_1-n_2)!}{(n-n_1-n_2)!}}_{=1} \cdot \underbrace{\frac{(n-n_1-n_2-n_3)!}{(n-n_1-n_2-n_3)!}}_{=1}$$

*Multiply Defn by three factors of unity*

$$= \frac{n!}{n_1!(n-n_1)!} \cdot \frac{(n-n_1)!}{n_2!(n-n_1-n_2)!} \cdot \frac{(n-n_1-n_2)!}{n_3!\underbrace{(n-n_1-n_2-n_3)!}_{=0!=1}} \cdot \underbrace{(n-n_1-n_2-n_3)!}_{=0!=1}$$

$$C^n_{n_1,n_2,n_3} = {}^nC_{n_1} \cdot {}^{n-n_1}C_{n_2} \cdot {}^{n-n_1-n_2}C_{n_3} = {}^nC_{n_1} \cdot {}^{n-n_1}C_{n_2}$$

$\underbrace{\phantom{xxxxxx}}$ n obj take $n_1$    $\underbrace{\phantom{xxxxxx}}$ n- $n_1$ obj remain take $n_2$    $\underbrace{\phantom{xxxxxx}}$ n- $n_1$ - $n_2$ obj remain take $n_3$

**1) Class Seating:** 70 Students seated in classroom 12 in each of 1st five rows & 10 in last row. Given that *order within a row is considered unimportant*

| | | | | | |
|---|---|---|---|---|---|
| Row #1 | 1 | 2 | 3 | .. | 12 |
| Row #2 | 1 | 2 | 3 | .. | 12 |
| Row #3 | 1 | 2 | 3 | .. | 12 |
| Row #4 | 1 | 2 | 3 | .. | 12 |
| Row #5 | 1 | 2 | 3 | .. | 12 |
| Row #6 | 1 | 2 | 3 | .. | 10 |

Direct Applic of Multinomial $\{n_1,n_2,n_3,n_4,n_5,n_6\} = \{12,12,12,12,12,10\}$

Binomial Decomp of Multinomial $\quad C^{70}_{12,12,12,12,12,10} = \dfrac{70!}{(12!)^5 (10!)^1}$

$$C^{70}_{12,12,12,12,12,10} = {}^{70}C_{12} \cdot {}^{70-12}C_{12} \cdot {}^{70-24}C_{12} \cdot {}^{70-36}C_{12} \cdot {}^{70-48}C_{12} \cdot \underbrace{{}^{70-60}C_{10}}_{=1} = 1.3 \times 10^{50}$$

**2) Football:** 25 Students into 3 categories 12- L 5-LB 8-DB

$$C^{25}_{12,5,8} = \frac{25!}{12!\,5!\,8!} = 6,692,786,100$$

***Combinatorial Explosions!***

The trinomial coefficient can conveniently be written as a product of three binomial coefficients by first choosing $n_1$ objects from n, then $n_2$ objects from the remaining $n-n_1$, and finally $n_3$ objects from the remaining $n-n_1-n_2=n_3$. The application of a trinomial coefficient to a problem may not always be obvious, but this decomposition shows that it may always be thought of as the product of two related binomials (the third one is always unity ${}^{n-n1-n2}C_{n3} = {}^{n3}C_{n3} =1$)

The example seating a class in a number of rows, where order *within a row is unimportant*, translates directly into a multinomial evaluation, as does the distribution of 25 members of a football team into three positions, where order within the position does not matter.

What if the order of seating within the row is important? Consider the simple case of just 22 students in two rows with 12 and 10 seats respectively. If order were not important we would compute as in the original example ${}^{22}C_{12}$ ${}^{22-12}C_{10} = {}^{22}C_{12} = 22!/(12!\ 10!) = 646,646$. However, when order is important, first permute 22 into the first 12 seats to obtain ${}^{22}P_{12} = 22!/(22-12)! = 22!/10!$; then permute the remaining 10 into 10 seats ${}^{10}P_{10} = 10!/(10-10)! = 10!$. Their product is ${}^{22}P_{12}\ {}^{10}P_{10} = (22!/10!)\ (10!) = 22! = 1.1240007\ 10^{21}$. We note that the partition into rows is completely irrelevant since we could just as well join them together into a single row of 22 seats and get the same answer ${}^{22}C_{12} = 22!$ .

## 2.3.4.2 Counting Examples

# Counting Examples

**Ex.1**    Committee of 2W & 3 M from{5W, 7M}   No restrictions    $\binom{5}{2}\cdot\binom{7}{3}=350$

$M_1M_2$ *Not Together;*   $M_1M_2^c$   $\binom{5}{2}\cdot\binom{5}{2}=100$   ;   $M_1^cM_2$   $\binom{5}{2}\cdot\binom{5}{2}=100$   ;   $M_1^cM_2^c$   $\binom{5}{3}\cdot\binom{5}{2}=100$    $=300$

---

**Ex.2**    Consider n =5 antennas , m=2 defective; n-m=5-2=3 OK
Arrange so that no 2 defective antennas are together when
placed among 3 good antennas

3 "good antennas"

$\wedge\ ^1\wedge\ ^2\wedge\ ^3\wedge$

Thus choose 2 positions from 4    $\binom{4}{2} \xrightarrow[\text{Generalize}]{} \binom{n-m+1}{m}$    "4" Available Positions $\wedge$
for 2 defective antennas

---

**Ex.3**    Distinct Assignments of 10
Police Officers: Streets: 5,
Station: 2 , Reserve 3:
$\dfrac{10!}{5!\,2!\,3!}$

Alternately,

remaining

$\binom{10}{5}\cdot\binom{5}{2}\cdot\binom{3}{3}=\dfrac{10!}{5!\,5!}\cdot\dfrac{5!}{2!\,3!}=\dfrac{10!}{5!\,2!\,3!}=2520$

---

**Ex.4a**    10 Children into 5-member; Teams A , B **do not** play one another    $\dfrac{10!}{5!\,5!}=252$

---

**Ex.4b**    Teams A , B do play one another    $\dfrac{1}{2!}\cdot\dfrac{10!}{5!\,5!}=126$    A,B equiv B,A
divide by Perm of
A,B =2!

**Choosing committees with and without restrictions:** In simple cases without restrictions, the answer may be written down directly in terms of combinations and products of combinations. In cases for which the restrictions are involved, it is wise to decompose the problem statements using set algebra to define ME and CE events as illustrated in the 1st panel. (It can be helpful to map out the event space with a tree as was done in Slide#2-23.)

**Defective Antenna Placement**: Place m=2 defective antennas between n-m good antennas without two defectives in one slot. Sketch identifies $3 + 1 = 4$ possible locations for defective antennas among 3 good ones leading to $^4C_2$ arrangements. Generalize to $^{n-m+1}C_m$ for n antennas with m defectives.

**Multinomial example** places 10 police officers into 3 groups – order not important within group.

**Choose two 5-member teams A and B** from 10 children. How many different teams are possible if (i) A, B are assigned to different leagues $L_1$ and $L_2$ and do not play against one another. Here the order A, B and B, A are considered different since they assign A and B to different leagues; *i.e.*, A playing in $L_1$ is different team choice than B playing in $L_1$.
(ii) A and B may play each other as in a playground. A, B and B, A are considered same match up. This is a subtle difference between *intermural* and *intramural* team selection.

## 2.3.4.3 Trinomial Expansion for n-Draws

### Trinomial Expansion for n-Draws

**Trinomial Expansion with n=2 draws:**

$$(x_1 + x_2 + x_3)^2 = \begin{pmatrix} 2 \\ 2\ 0\ 0 \end{pmatrix} x_1^2 \cdot x_2^0 \cdot x_3^0 + \begin{pmatrix} 2 \\ 0\ 2\ 0 \end{pmatrix} x_1^0 \cdot x_2^2 \cdot x_3^0 + \begin{pmatrix} 2 \\ 0\ 0\ 2 \end{pmatrix} x_1^0 \cdot x_2^0 \cdot x_3^2$$

$$+ \begin{pmatrix} 2 \\ 1\ 1\ 0 \end{pmatrix} x_1^1 \cdot x_2^1 \cdot x_3^0 + \begin{pmatrix} 2 \\ 1\ 0\ 1 \end{pmatrix} x_1^1 \cdot x_2^0 \cdot x_3^1 + \begin{pmatrix} 2 \\ 0\ 1\ 1 \end{pmatrix} x_1^0 \cdot x_2^1 \cdot x_3^1$$

$$= x_1^2 + x_2^2 + x_3^2 + 2x_1 x_2 + 2x_1 x_3 + 2x_2 x_3$$

Number of terms in trinomial expansion is $\quad {}^3\!/\!_2 = {}^{3+(2-1)}C_2 = (4*3)/2! = 6$

**Trinomial Expansion with n-draws:**

$$(x_1 + x_2 + x_3)^n = \begin{pmatrix} n \\ n\ 0\ 0 \end{pmatrix} x_1^n \cdot x_2^0 \cdot x_3^0 + \begin{pmatrix} n \\ 0\ n\ 0 \end{pmatrix} x_1^0 \cdot x_2^n \cdot x_3^0 + \begin{pmatrix} n \\ 0\ 0\ n \end{pmatrix} x_1^0 \cdot x_2^0 \cdot x_3^n$$

$$+ \begin{pmatrix} n \\ n-1\ 1\ 0 \end{pmatrix} x_1^{n-1} \cdot x_2^1 \cdot x_3^0 + \begin{pmatrix} n \\ 1\ 0\ n-1 \end{pmatrix} x_1^1 \cdot x_2^0 \cdot x_3^{n-1} + \begin{pmatrix} n \\ 0\ n-1\ 1 \end{pmatrix} x_1^0 \cdot x_2^{n-1} \cdot x_3^1$$

$$+ \text{(many more terms!!)}$$

Number of terms in general trinomial expansion is $\quad {}^3\!/\!_n = {}^{3+(n-1)}C_n = {}^{n+2}C_n = (n+2)(n+1)/2$

(See Slide#2-31 *ff.* for combinations with replacement "slash- ${}^nC_k$" )

Here is an explicit evaluation of the trinomial expansion for n=2. Notice that j,k,l each range over indices 0,1,2, but not freely since their sum must always be n=2.

In the general trinomial case $(p_1+p_2+p_3)^n$, the coefficient values (multiplicities) are given by the multinomial values ${}^nC_{i1,i2,i3}$ with constraint $i_1+i_2+i_3 = 3$. The *number of distinct terms in the expansion* (groupings) is given by the number of integer solutions to the sum constraint equation $i_1+i_2+i_3 = 3$ solution is slash- ${}^3C_n = {}^{3+n-1}C_n = {}^{n+2}C_n = (n+2)(n+1)/2$ which is equivalent to the number solutions to the n- balls and r-bins problem, *viz.*, ${}^{r+n-1}C_{r-1}$ for r=3.

Clearly, the expansion of a trinomial for larger values of n=2, 3, 4, ... becomes increasingly more difficult to write out explicitly without the use of a computer algorithm because the number of terms increases quadratically $(n+2)(n+1)/2 \sim n^2/2$.

## 2.4 Repeated Trials

---

# Repeated Trials

- Counting with Replacement
- Integer Equation Solutions
- Balls and Bins
- Perms/Combos/Replacement/Trees
- Potpourri Summaries
- Repeated Trials/Multinomials
- Multinomial Expansions and Distributions

---

Understanding how to enumerate all possible outcomes correctly is an essential element of discrete probability. If all outcomes are equally likely, then the count of outcomes associated with a given event divided by the total number of outcomes yields the probability of that event. There are situations in which the *selected objects are not replaced* and these are handled using permutations and combinations to enumerate the outcomes. In other situations the objects are replaced after each selection, so that we can repeat these trials indefinitely; we will need to extend our computational methods to include *permutations and combinations with replacement*. Algebraic set theory, Venn diagrams, and trees help us understand these nuances in counting.

We shall also revisit the binomial, trinomial and multinomial expansions which relate to repeated trials with 2, 3, and r possible outcomes, respectively, and visualize these outcomes as the branches emanating from the nodes of an associated tree.

There is inevitably a conflation of these various concepts involving permutations and combinations *with and without replacement*, tree diagrams, integer equation solutions, balls and bins, "when do we multiply?", "when do we add?" and we will use a few "potpourri slides" to make the appropriate connections and try to disentangle these often confusing concepts.

## 2.4.1 Counting with Replacement

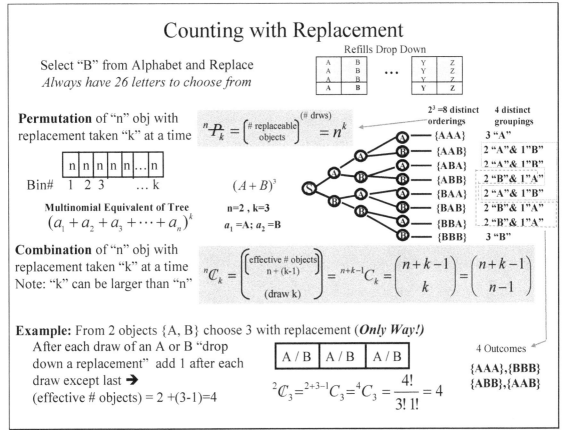

Counting permutations and combinations *with replacement* is analogous to the operation of a candy machine in which a new object drops down to replace the one just purchased, thus giving the same number of choices for each customer. A tree describing k draws from n objects gives all permutations by simply adding n sub-branches at each draw, yielding a total of $n^k$ permutations (outcomes); in general, the same tree has fewer distinct groups than permutations at its output. In order to properly enumerate the number distinct groups output by the tree, the "effective" number of objects must be increased from n to n+(k-1) in order to account for the (k-1) drop-down replacements made prior to last draw.

**Permutation of n objects taken k at a time with replacement:** Because of object replacement after each draw, there are always n possible choices, thus yielding $n*n*n...*n = n^k$; this is written using the notation "slash-$^nP_k$" with a "slash" across the permutation symbol as shown. The case n=2, k=3 represents 3 draws from 2 objects, which is impossible except in the case of *two replaceable objects* {A, B}. The figure shows the slash-$^2P_3 = 2^3 = 8$ permutations that result.

**Combination of n objects taken k at a time with replacement:** For n=2, k=3, "taking 3 from 2" or $^2C_3$ normally does not make sense. However, "taking 3 from 2" *with replacement* or "slash-$^nC_k$" does, because each draw (except the last) drops down an identical item and hence the "effective number of items" to choose from becomes n+(k-1) and slash-$^nC_k = {}^{n+(k-1)}C_k$. The tree shows that there are 4 distinct groupings {3A, 3B, 2A1B, 1A2B}, which is exactly the number of combinations with replacement given by the general formula for the case n=2, k=3: slash-$^2C_3 = {}^{2+(3-1)}C_3 = {}^4C_3 = 4$.

## 2.4.2 Integer Solutions of Equations (n Balls into r Bins )

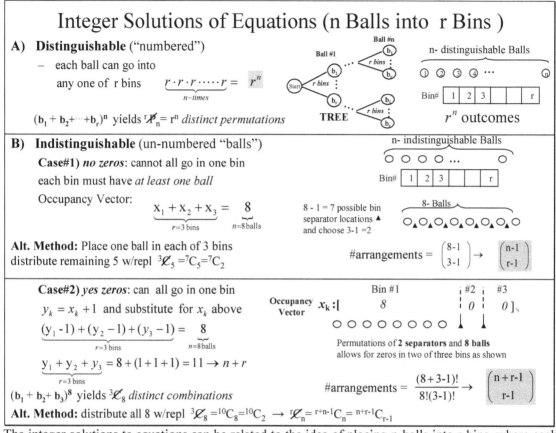

The integer solutions to equations can be related to the idea of placing n-balls into r-bins, where each bin can accept any number of balls. The result can also be interpreted in terms of the ideas of permutations and combinations with replacement (see "Alt. Method" on this slide; also Slide# 2-33).

**A) If the balls are numbered (distinct)** then each of the n distinct balls is drawn in turn and can be placed into any one of the r bins. This is analogous to permutations with replacement where the *bins serve as the replaceable object* and the *ball serves as the "draw"*. The result is just slash-$^{r}P_n = r^n$ possible arrangements of n-balls in the r-bins as illustrated by the tree and the multinomial algebraic.

**B) If the balls are un-numbered (not distinct)** then each of the n identical balls can again be placed in any bin. However, the number of distinct arrangements of *un-numbered* balls in bins is much smaller than in A). The analysis can be framed in terms of an occupancy *vector* whose components represent the number of balls in each of the r-bins; the sum of occupancy numbers equals the total number of balls n. For the specific case n=8, r =3 this yields an equation: $x_1+x_2+x_3=8$ with occupation vector $[x_1, x_2, x_3]$. Consider the placement of (3-1)=2 **bin separators** "among" the 8 balls in such a way that none of the three bins is empty (no zero components). We see that there are a total of 8-1 possible separator locations between the 8 balls; outside positions yield empty bins. The two separators form 3 bins if placed in any one of the 7 positions, *i.e.*, there are $^{8-1}C_{3-1}$ possible arrangements (in general: $^{n-1}C_{r-1}$). The case in which the occupation number $y_k$ can be zero yields a slightly different result. It is obtained by the algebraic substitution $y_k = x_k + 1$, which allows the $x_k$ to be zero but keeps $y_k \geq 1$. Applying the previous result for non-zero occupation to $y_k$ yields $^{(8+3)-1}C_{3-1}$, which, upon mapping 8→n and 3→r, yields the general result $^{n+r-1}C_{r-1}$. Also, the number of distinct tree groups or distinct terms of the trinomial give the same result as shown at the bottom of this slide.

## 2.4.2.1 Relation of Balls and Bins to Counting with Replacement

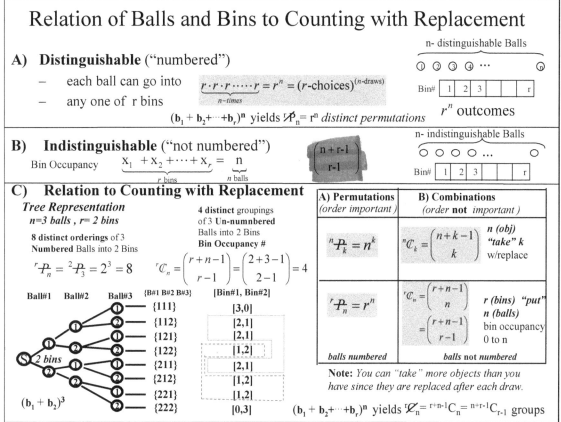

# Relation of Balls and Bins to Counting with Replacement

**A) Distinguishable** ("numbered")

– each ball can go into
– any one of r bins

$$\underbrace{r \cdot r \cdot r \cdots \cdot r}_{n-times} = r^n = (r\text{-choices})^{(n-draws)}$$

$(b_1 + b_2 + \cdots + b_r)^n$ yields $^r\!P_n = r^n$ distinct permutations

n- distinguishable Balls

① ② ③ ④ ⋯  ⓝ

Bin# | 1 | 2 | 3 | | | r

$r^n$ outcomes

**B) Indistinguishable** ("not numbered")

Bin Occupancy $\quad \underbrace{x_1 + x_2 + \cdots + x_r}_{r\ bins} = \underbrace{n}_{n\ balls}$

$\dbinom{n+r-1}{r-1}$

n- indistinguishable Balls

○ ○ ○ ○ ⋯  ○

Bin# | 1 | 2 | 3 | | | r

**C) Relation to Counting with Replacement**

*Tree Representation*
*n=3 balls , r= 2 bins*

**8 distinct orderings** of 3
**Numbered** Balls into 2 Bins

$$^r\!P_n = {}^2\!P_3 = 2^3 = 8$$

**4 distinct** groupings
of 3 **Un-numbered**
Balls into 2 Bins
**Bin Occupancy #**

$$^r\!C_n = \dbinom{r+n-1}{r-1} = \dbinom{2+3-1}{2-1} = 4$$

| | **A) Permutations** *(order important )* | **B) Combinations** *(order **not** important )* | |
|---|---|---|---|
| | $^n\!P_k = n^k$ | $^n\!C_k = \dbinom{n+k-1}{k}$ | n (obj) "take" k w/replace |
| | $^r\!P_n = r^n$ | $^r\!C_n = \dbinom{r+n-1}{n}$ $= \dbinom{r+n-1}{r-1}$ | r (bins) "put" n (balls) bin occupancy 0 to n |
| | *balls numbered* | *balls not numbered* | |

| Ball#1 | Ball#2 | Ball#3 | {B#1 B#2 B#3} | [Bin#1, Bin#2] |
|---|---|---|---|---|
| | | ① | {111} | [3,0] |
| | | ② | {112} | [2,1] |
| | | ① | {121} | [2,1] |
| | 2 bins | ② | {122} | [1,2] |
| | | ① | {211} | [2,1] |
| | | ② | {212} | [1,2] |
| | | ① | {221} | [1,2] |
| $(b_1 + b_2)^3$ | | ② | {222} | [0,3] |

**Note:** *You can "take" more objects than you have since they are replaced after each draw.*

$(b_1 + b_2 + \cdots + b_r)^n$ yields $^r\mathbb{C}_n = {}^{r+n-1}C_n = {}^{n+r-1}C_{r-1}$ groups

**The first panel A)** shows the result for distinct numbered balls: Each ball in turn "selects" one of r bins; the bins are the "replaceable" element meaning that each ball has the same choice of r bins. The tree (not shown) always has r branches emanating from each node.

**The second panel B)** has been previously discussed in the balls and bins slide and is repeated here for convenience.

**The last panel C)** illustrates the relationship (mapping) between the "balls and bins analysis" and the "combinations with replacement analysis." The tree representation of n=3 balls into r =2 bins shows the 2-bin choice emanating from each node and labels the output by the sequence of bin choices. For indistinct balls the specification of which ball is in a given bin is superfluous. Hence, the output state labels {112}, {121}, and {211} actually represent the same state, namely, the placement of 2 indistinct balls in bin #1 and 1 ball in bin#2. Thus, the 2-element "bin occupation# vector" [2,1] suffices for all three and is shown as a single state dashed outline (green). Similarly, [1,2] represents the state with 1 ball in bin #1 and 2 indistinct balls in bin#2 solid-outlined (red). In all, there are 4 bin occupation vectors {[2,1], [1,2], [3,0], [0,3]} shown in the tree. The number of distinct outputs is computed directly as a combination with replacement by considering 3 draws from 2 bins with "bin replacement", *viz.,* slash-$^2C_3 = {}^{2+(3-1)}C_3 = {}^4C_3 = 4$.

In the "n-balls r-bins problem" the r-bins are the "replaceable objects"; thus we can apply combinations with replacement directly as the statement "r-bins choose n-balls" which translates to slash-$^rC_n$, and yields the balls and bins result, *viz.* ,

$$\text{slash-}^rC_n = {}^{r+(n-1)}C_n = (r+n-1)! \, / \, \{n! \, (r-1)! \, \} = {}^{n+r-1}C_{r-1}.$$

## 2.4.3 Potpourri: Perms/Combos/Replacement/Trees - 2 Balls, 3 Bins

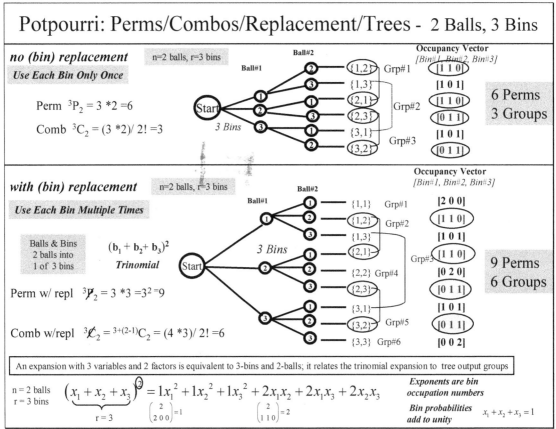

This slide and the next give further insight into the relationships between "balls and bins analysis," "trees," and "permutations and combinations". We also show how the tree output can be interpreted in terms of a multinomial expansion. In this slide we consider n=2 balls as "the draws" into r=3 bins, both *with* and *without* "bin" replacement; in the next we take n=3 balls as "the draws" into r=2 bins.

**The upper tree** "without bin replacement" allows 3 bin choices for the 1st ball but only 2 choices for the 2nd ball and yields 6 permutation outputs and just 3 groups as shown. A computation using *permutations and combinations without replacement* yields precisely the same results derived from the explicit tree.

**The lower tree** "with bin replacement" allows 3 bin choices for the 1st ball and also 3 choices for the 2nd ball and yields 9 permutation outputs and just 6 groups as shown. A computation using *permutations and combinations with replacement* yields precisely the same results derived from the explicit tree.

**In both trees**, specification of a **3-component bin occupation#**, *e.g.*, [0 1 1] provides a unique label for the group of two states {2,3} and {3,2} in which ball#1 occupies bin#2 and ball#2 occupies bin#3 and *vice versa*. Since the balls are indistinguishable the two states are identical.

An **interesting interpretation of balls and bins** is found by expanding the trinomial $(x_1 + x_2 + x_3)^2$, where the variables are the probabilities for choosing the 3 bins and add to unity; the exponent 2 represents the number of balls to be placed in bins. The algebraic expansion of the trinomial yields 6 distinct probability terms of the form $^2C_{i,j,k} \cdot x_1^i \cdot x_2^j \cdot x_3^k$, where the *exponents* of the "x"s represent the *bin occupation numbers* and the trinomial coefficients $^2C_{i,j,k}$ are the relative weights (or multiplicities) of the trinomial tree groupings. For example, the term $^2C_{110} \cdot x_1^1 \cdot x_2^1 \cdot x_3^0$ represents one ball in each of bins #1 and #2 and none in bin#3 and its multiplicity $^2C_{110} = 2$ corresponds the two members of tree group#2.

## 2.4.3.1 Potpourri: Perms/Combos/Replacement/Trees - 3 Balls, 2 Bins

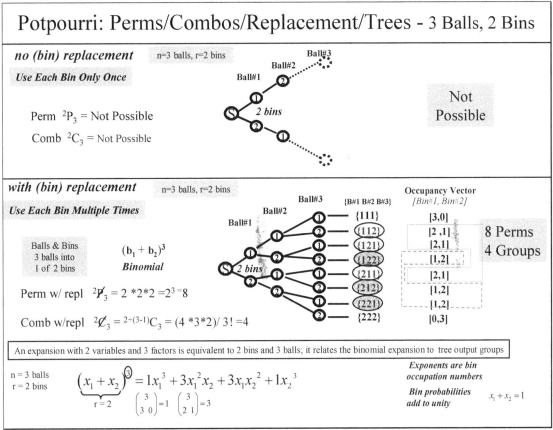

In this companion slide we consider n=3 balls as "draws" into r=2 bins both with and without replacement.

**The upper tree** "without bin replacement" allows 2 bin choices for the 1st ball, but only 1 choice for the 2nd ball and no choice for the 3rd ball; it illustrates the obvious fact that we cannot choose (draw) 3 bins from 2 when there is no bin replacement!

**The lower tree** "with bin replacement" allows 2 bin choices for the 1st ball and now allows the same 2 choices for the 2nd ball and again for the 3rd ball. It thus yields $2*2*2 = 2^3=8$ permutation outputs, but just 4 distinct groups as shown in the tree. The tree output is labeled by the bin choices for each ball; in this scheme the outcomes {1 1 2}, {1 2 1} and {2 1 1} all belong to the same group (two balls in bin #1 and one in bin#2). All three states (members of this "green" group) may thus be specified by the same 2-component bin occupation # "[2 1]" since each member has exactly 2 balls in bin#1 and 1 ball in bin#2. There are also three identical members to the "red" group labeled [1,2] and finally two singleton groups [3,0] and [0,3] leading to 4 distinct groups in all. Computation using the formulas for permutations and combinations with replacement yields precisely the same results obtained by direct inspection of the tree output.

An interpretation of balls and bins is found in this case by expanding the binomial $(x_1 + x_2)^3$, where now the two variables are the probabilities for choosing the 2 bins and add to unity; the exponent 3 represents the number of balls to be placed in bins. The algebraic expansion of the binomial yields 4 distinct probability terms of the form $^3C_{i,j} \cdot x_1^i \cdot x_2^j$, where the *exponents* of the "x"s represent the *bin occupation numbers* and the binomial coefficients $^3C_{i,j}$ are the relative weights (or multiplicities) of the groups that appear in the binomial tree. For example, the term $1 \cdot x_1^3$ represents all three balls in bin #1 and its multiplicity of "1" corresponds to the top singleton in the tree. Similarly, the term $3 \cdot x_1^2 x_2$ represents two balls in bin #1 and one ball in bin#2 and its coefficient $^3C_{21} = 3$ corresponds to the three members of that group in the tree. This can be generalized to give the distribution of any number balls into any number of bins.

## 2.4.3.2 Potpourri: Repeated Trials/Multinomials

# Potpourri: Repeated Trials/Multinomials

**Binomial**

$$(a+b)^3 = \sum_{k=0}^{3}\binom{3}{k}a^k b^{3-k}$$

$\#groups = {}^2\mathscr{C}_3 = {}^{2+3-1}C_3 = {}^4C_1 = 4$

$\#permutations \quad {}^2\mathscr{P}_3 = 2^3 = 8$

**n=3 Binomial PMF**

$p_X(x)$

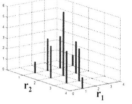

**Tree Expansion/ Grouping**   a=b=1

$$(2)^3 = \binom{3}{0}+\binom{3}{1}+\binom{3}{2}+\binom{3}{3}$$

**Random Variable PMF**

$$(p+q)^3 = \binom{3}{0}p^0 q^3 + \binom{3}{1}p^1 q^2 + \binom{3}{2}p^2 q^1 + \binom{3}{3}p^3 q^0$$

a=p ; b=q ; p+q=1

| Repeated Indep Trials | #ways $2^3=8=$ | 1 | 3 | 3 | 1 |
|---|---|---|---|---|---|
| | #succ | 0 | 1 | 2 | 3 |
| | #fail | 3 | 2 | 1 | 0 |

**Trinomial**

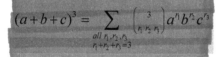

$$(a+b+c)^3 = \sum_{\substack{all\ r_1,r_2,r_3 \\ r_1+r_2+r_3=3}}\binom{3}{r_1\ r_2\ r_3}a^{r_1}b^{r_2}c^{r_3}$$

$\#groups = {}^3\mathscr{C}_3 = {}^{3+(3-1)}C_3 = (5*4*3)/3! = 10$ terms

$\#permutations = {}^3\mathscr{P}_3 = 3*3*3 = 3^3 = 27$ terms

**Tree Expansion/ Grouping**   a=b=c=1

$$(3)^3 = 27 = \underbrace{\binom{3}{0\,0\,3}+\binom{3}{0\,1\,2}+\cdots+\binom{3}{0\,3\,0}+\cdots+\binom{3}{3\,0\,0}}_{10\ terms\ or\ ``groups"}$$

**Random Variable PMF**
a=p ; b=q ; c=r:
p+q+r =1

$$1=(p+q+r)^3=\binom{3}{0\,0\,3}p^0 q^0 r^3 +\binom{3}{0\,1\,2}p^0 q^1 r^2 +\cdots+\binom{3}{3\,0\,0}p^3 q^0 r^0$$

**Repeated Indep Trials**

*[Bin#1, Bin#2, Bin#3]*
**Occupancy Vectors**

Later on we will analyze repeated independent trials which involve precisely the "draws with replacement" that we have been discussing. Such trials each have an equal probability of success p and failure q =1- p. When there are many trials it makes sense to talk about the distribution of outcomes which is called a probability distribution. One such distribution is the binomial distribution which is a direct consequence of the binomial expansion and is plotted explicitly in the figure. The terms in the binomial expansion when normalized to unity are precisely the probabilities associated with the binomial distribution of outcomes.

The binomial expansion can be "normalized" by either setting a=b=1 and then dividing by $(1+1)^n$ to yield unity on the LHS and the distribution function on the right. Alternately, substituting a=p, b=q=1-p in the binomial expansion yields $(p+1-p)^n=1$ again on the LHS and the distribution on the RHS. The first method holds for p=q=1/2 while the second method holds for all choices of the single trial probability of success p.

Similar remarks hold for the trinomial distribution shown in the bottom panel; since the exponents satisfy $r_1 + r_2 + r_3 =3$, the values of 10 multinomial coefficients may be plotted in the $(r_1, r_2)$ –plane. (Note: Plot shown is for the case p = q = r = 1/3. All coefficients in the plot must be divided by 27 and the $(r_1, r_2)$ –values must be decreased by 1). It should also be pointed out that as the number of trials n becomes large the distribution approaches a Bivariate Gaussian.

# 3 Numerical Assignment of Probability

### *3.1 Probability Models and Computation Techniques*

# Probability Models and Computation Techniques

1.  Axioms

2.  Formulations of Probability

3.  Compound Events and Set Algebra

4.  Adding Probabilities: Inclusion / Exclusion, CE & ME

5.  Application of Venn Diagrams and Trees

6.  Committees, Urns, and Pairings

7.  Solution to Integer Equations

8.  Man-Hat Matching Problem

Probability is based on a small set of axioms that determine how numerical values are assigned to the occurrence of random events. We are free to assign probabilities on the basis of a theoretical model, experimental observations, Bayesian belief methods, or even some *ad hoc* or subjective reasoning, so long as the axioms are satisfied. The axioms also require probability to be a number in the interval [0, 1]; although obvious, it bears mention that *probability may be zero, but it is never negative.*

To be valid, an event space must consist of ME and CE elements. Evaluating probabilities for compound events is naturally developed *via* set algebra using unions, intersections, and complements composed of elements belonging to the "event space." Further manipulation using DeMorgan's laws, and inclusion/exclusion expansions usually suffices to solve the problem, but often the solution is made more transparent by formulating it in a new event space. Visualization aides such as Venn diagrams, trees, coordinate plots, and other appropriate graphical constructs are often helpful both in formulating and validating results.

There is a large variety of problem types, ranging from simple committee formation and classic urn problems to more intricate problems that involve object pairing, integer constraint equations, or special matching conditions. It is wise to apply a number of different computation methods to difficult problems in order to guard against possible misinterpretation as well as to gain additional insight. The "Man-Hat" matching problem is a case in point; a multi-pronged approach using set algebra, tables, trees, and Venn diagrams, leads to more complete understanding of how two different event space formulations complement one another and yield insights that otherwise would be lost.

## 3.2 *Formulations of Probability*

---

# Formulations of Probability

**Axioms of Probability**

*i) Non -negative  P(E) ≥ 0*

*ii) Normalization  P(S)=1     S= Entire Sample Space*

*iii) Addition of Probabilities for ME Events  $E_1E_2=\phi$; $P(E_1 \cup E_2)=P(E_1)+P(E_2)$*

  Note from Axioms ii) and iii) Probability is bound between 0 and 1

  $1=P(S)=P(E \cup E^c)=P(E)+P(E^c) \implies P(E)=1 - P(E^c) \le 1$          $0 \le P(E) \le 1$


**Numerical Assignment of Probability**

1) *Classical Definition* - Counting Theoretical Models  (Equally likely outcomes)

2) *Relative Frequency of Occurrence* - Observeration of Finite Number of Experiments

3) *Bayesian Belief*:  Based on Measurement and *a priori (belief) assignments*

4) *Subjective Belief*: Based on "generally accepted" or  Personal beliefs.

---

**Axioms of Probability**
 i)  Probability of an event is never negative; it can be zero or positive.
 ii)  Probability is normalized to unity so the probability of any given event is less than or equal to 1.
 iii) Probabilities of non-intersecting events simply add
Probability of a given event is a number in the interval [0,1]
Probability may be formulated in terms of "equally likely outcomes," "observed frequency-of-occurrence," "a Bayesian measure of belief," or "by some *ad hoc* (subjective) numerical assignment." We shall briefly discuss each of these formulations on the following four slides.

## 3.2.1 Classical Definition of Probability

# 1) Classical Definition of Probability

Develop a *theoretical model* of equally likely outcomes and count

$$S = \{s_1, s_2, s_3, \cdots, s_n\}$$

"Atomic" outcomes - Equally Likely because by definition they are point outcomes with no distinguishing characteristics that might make one more likely than another

$$s_i \cap s_j = \varphi \quad all \ i \neq j$$

$$P(s_i) = \frac{1}{n} \ ; \quad all \ i$$

Additivity Axiom for non intersecting events yields:

$$P(s_1 \cup s_2) = \frac{1}{n} + \frac{1}{n} = \frac{2}{n} \ etc.$$

$$P(S) = P\left(\bigcup_{i=1}^{n} s_i\right) = \sum_{i=1}^{n} P(s_i) = \sum_{i=1}^{n} \frac{1}{n} = \frac{n}{n} = 1$$

Classical Defn:
$$P(E) = \frac{n(E)}{n(S)} = \frac{(\#\,\text{eq. likely outcome in E})}{(\text{Total \# outcomes in S})}$$

Example: Pair of 6-sided Dice

Sum Event $E = d_1 + d_2$

Total Outcomes $n(S) = 6^2 = 36$

| E=d1+d2 | n(E) | P(E) |
|---------|------|------|
| 2, 12   | 1    | 1/36 |
| 3, 11   | 2    | 2/36 |
| 4, 10   | 3    | 3/36 |
| 5, 9    | 4    | 4/36 |
| 6, 8    | 5    | 5/36 |
| 7       | 6    | 6/36 |

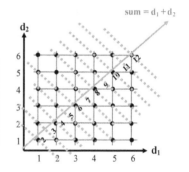

The random throws of a pair of dice defines a sample space with equally likely atomic events parameterized by coordinates $(d_1, d_2)$; thus simple counting is used to define a theoretical probability model.

Imposing game rules as in "craps" defines complex events and the theoretical probability model for "craps" is determined by set algebra of these events. Sum and difference coordinates are the most natural way to express the outcomes.

The $(d_1, d_2)$ coordinate representation (on bottom right) for a pair of 6-sided dice has 36 *equally likely outcomes* represented by "dots", so they each have probability of 1/36. Overlaid on the figure is the sum axis $s = d_1 + d_2$ at +45 degrees as well as a set of 11 dashed lines running diagonally at -45 degrees passing through grid points having the same sum value. The table values for sum probabilities can be read directly off this plot by adding the equally likely "dot" events corresponding to a given sum and dividing by 36. Thus, for the sum $s = d_1 + d_2 = 4$ (or 10) we count diagonally along the corresponding dashed line to find three events and obtain the probability of 3/36 as listed in the table.

Note that the sum s=7 diagonal line is the symmetry or "folding" line about which the other sums are symmetric; thus column#1 of the table specifies only pairs with equal probabilities {2,12}, {3,11}, {4,10}, {5,9}, {6,8} and the unique value for {7}.

## 3.2.2 Relative Frequency of Occurrence

# 2) Relative Frequency of Occurrence

**Relative Frequency Defn** - Experimental Results from a Large Number of Trials

$$P(E) = \lim_{N\to\infty} \frac{n_E(N)}{N} = \lim_{N\to\infty} \frac{(\#\text{ times E occurs})}{(\text{Total } \#\text{ trials})}$$

**Pros:**

1) Based on **Experimental Fact** - Repeated Experiments

2) Not limited to **equally likely outcomes**

3) Abstract axioms not imposed *e.g.* **"fair coin"**

**Cons:**

1) Extremely **rare events** may not occur in limited # of trials

2) Outcomes of repeated trials **may not be independent**

3) **Repeated Trials** may not be practical

4) **Limit** may not exist (may oscillate)

**Ex.:**

Flip coin n=1000 and find experimentally $n_H$ = 727 heads

compute: $P(n_H) = 727/1000 = .727 \Longrightarrow P(n_T) = 1 - P(n_H) = .273$

conclude: not a fair coin

Experimental results for coin tosses, dice throws, *etc.*, define an empirical probability. Such a model is subject to the conditions of the particular experiment and may not be repeatable; moreover, rare events may not occur in any given sample. However, it has the advantage that we do not have to assume a fair coin, but rather let the experiment determine the fairness of the coin.

One may consider gathering empirical data in two complementary ways

**(i) A Time Series:** Collecting a series of measurements in time (at a fixed point in space) and assuming that the underlying stochastic process we are sampling does not change in time, or

**(ii) An Ensemble:** Collecting measurements over a region of space at single instant of time ("snapshot") and assuming that the underlying stochastic process we are sampling does not change with location.

An **ergodic process** is one for which statistical averages over time or over space yield the same results; more specifically, this means that analysis of *time series data* yields the same results as *ensemble data*. However, if the underlying stochastic process changes over time or space, then the relative frequency-of-occurrence calculated from a time series may well be different from that calculated from an ensemble. Thus, although frequency-of-occurrence is a natural probability model, it has inherent limitations because it may depend upon when, where, and at what rate the data is captured.

### 3.2.3 Bayesian Belief

# 3) Bayesian Belief

- **Bayes' Rule:** *Measurement Update  (~Kalman Filter)*
  - Two events {0, 1} states of system
  - *a priori* prob P(0)=P(1)=1/2
  - Two Measurement  Types {$M_0$, $M_1$}

    $\rightarrow$ give credence to existence of states{0,1}
  - *a posteriori* probability obtained

    $\rightarrow$ from Bayes meas. update
  - Measurement either *increases or decreases* "belief"  in state "0" or "1"
  - *In the absence of any measurements you can only assign the **a priori** beliefs which in this case are equal P(0)=P(1)=1/2*
  - Multiple measurements incrementally increase/decrease believability
  - Belief that a "1" was sent ***given a measurement $M_1$*** is the probability P(1| $M_1$)
  - A measurement $M_1$ ***given a "1" was sent*** is the inverse probability P($M_1$|1) with 1& $M_1$ interchanged
  - If "1" is sent the ***measurement $M_1$ is an effect;*** however if "$M_1$" is measured it ***does not cause the "1",*** it only increases belief that "1" was sent.
  - ***Thus it is best to think in terms of a correlation between the State "1" and the Measurement "$M_1$" rather than cause & effect.***

$$P(1\,|\,M_1) = \frac{P(M_1\,|\,1)\cdot P(1)}{P(M_1)}$$

$$P(M_1) = \underbrace{P(M_1\,|\,0)}_{\substack{\text{False Positive}\\\text{Measurement}\\\text{Statistic}}}\cdot \underbrace{P(0)}_{\text{a priori}} + \underbrace{P(M_1\,|\,1)}_{\substack{\text{Positive}\\\text{Measurement}\\\text{Statistic}}}\cdot \underbrace{P(1)}_{\text{a priori}}$$

Bayesian belief probability concepts are best understood by considering what happens when a binary digital signal is transmitted and then detected at by a remote observer. Either on theoretical grounds, or from previous data analysis we determine that "0"s and "1"s are equally likely, so that prior to making any measurement, the probabilities are equal P(0) = P(1) = 1/2. If a digital "1" is in fact transmitted, the observer continues to believe the digit is either a "0" or a "1" with equal probability; only a direct measurement $M_1$ declaring the digit a "1" or $M_0$ declaring it a "0" will change his *belief*. The so-called Bayesian measurement update changes the observer's belief (that the digit is a "1") from its *a priori* value of 1/2 to an *a posteriori* value P(1|$M_1$) >1/2 for measurement $M_1$ or to P(1|$M_0$) < 1/2 for measurement $M_0$.

Bayes' rule computes the updated probability P(1|$M_1$) by explicitly using pre-established detection probabilities P($M_1$|1) and P($M_1$|0) according to the formula given on the slide. The measurement statistic P($M_1$|1) is the probability that a measurement "$M_1$" is made  given that a "1" was sent, while P($M_1$|0) is the probability that a measurement "$M_1$" is made given that a "0" was sent (false positive.) Corresponding correct and false positive measurement statistics for $M_0$ are P($M_0$|0) and P($M_0$|1) respectively. Note that for a "good" detector, the correct detection statistics are usually close to unity and the false positives are very small.

The Bayes update is also known as "inverse probability" since the update expression P(1|$M_1$) is literally the measurement statistic P($M_1$|1) with its variables reversed. However, there is a more subtle issue here because the mathematical reversal actually inverts cause and effect. The existence of digit "1" causes the measurement $M_1$ by the detector so P($M_1$|1) is cause and effect; however, in P(1|$M_1$), the existence of $M_1$ does not in any sense cause a digital "1". Therein lays the philosophical issue with Bayes' rule. No matter, Bayes' rule is mathematically correct, very useful, and easily proved using the definition of conditional probability.

## 3.2.4 Subjective Belief and Consistency

# 4) Subjective Belief and Consistency

Two different subjective beliefs as to the probability of rain in the next two days are given in the following table

| Rain Events | Choice #1 | Choice#2 |
|---|---|---|
| Day 1 | $P(D_1) = .30$ | $P(D_1) = .30$ |
| Day 2 | $P(D_2) = .40$ | $P(D_2) = .40$ |
| Both $D_1$ & $D_2$ | $P(D_1D_2) = .20$ | $P(D_1D_2) = .10$ |
| Either $D_1$ or $D_2$ | $P(D_1 U D_2) = .60$ | $P(D_1 U D_2) = .60$ |

Since this is purely subjective, we could put it to a vote.

However, before doing that check to see they satisfy the three Axioms of Probability Assignment

Later we shall show that the axioms for intersecting events must satisfy the following formula:

$$P(D_1 \cup D_2) = P(D_1) + P(D_2) - P(D_1 D_2) \longleftarrow \text{Subtract out double counting}$$

Choice#1  $.60 \overset{?}{=} .30 + .40 - .20 = .50$  *No!*

Choice#2  $.60 \overset{?}{=} .30 + .40 - .10 = .60$  *Yes!*

Clearly Choice#2 has consistency going for it … if nothing else!

Subjective probability is nothing more than an opinion based on subjective data and/or intuition. Such guesses can be weeded out by making sure that they are at least consistent with the basic axioms of probability. Here we see that Choice#1 must be rejected because it does not satisfy the basic axioms of probability. Choice#2 is consistent, but has little else to recommend it.

## 3.3 *Probability for Compound Events*

# Probability of Compound Events

**Compound Events:** Aggregates of atomic events formed by unions and intersections

**Example #1:** 2 marbles selected from bag containing {**4 W, 6 R**} ; compound event E={ 1W 1R} (*order unimportant*)

**"unordered sample space"** $\qquad P(E) = P(1W,1R) = \dfrac{n(E)}{n(S)} = \dfrac{^4C_1 \cdot {}^6C_1}{^{10}C_2} = \dfrac{24}{45}$

**"ordered sample space"** $\qquad P(1W,1R) = P(\{WR\} \cup \{RW\}) = P(WR) + P(RW) = \underbrace{\dfrac{4}{10} \cdot \dfrac{6}{9}}_{\substack{\text{White First}}} + \underbrace{\dfrac{6}{10} \cdot \dfrac{4}{9}}_{\substack{\text{Red First}}} = \dfrac{48}{90}$

$\underbrace{\phantom{P(\{WR\} \cup \{RW\})}}_{\substack{\text{ME Events}\\ \text{Ordered Sample Space}}}$

---

**Example #2:** Random Integer $\beta \in [100,499]$ ; $\beta = IJK$ ; where, $I = 1,\cdots,4$ ; $J,K = 0,\cdots,9$

**a) $E_1$** $\beta$ contains at least one digit "1"

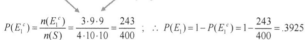

$A:\{"1"JK\}$ ; $B\{I"1"K\}$ ; $C\{IJ"1"\}$ ; $E_1 = A \cup B \cup C$

Events satisfying criterion can be composed from A,B,C $\quad E_1^c = (A \cup B \cup C)^c \underset{DeMorgan}{=} A^c B^c C^c \longleftarrow$ 3rd digit not "1"

2nd digit not "1"

1st digit not "1"

3 ways digit not "1" $\qquad$ 9 ways digit not "1"

$P(E_1^c) = \dfrac{n(E_1^c)}{n(S)} = \dfrac{3 \cdot 9 \cdot 9}{4 \cdot 10 \cdot 10} = \dfrac{243}{400}$ ; $\therefore\ P(E_1) = 1 - P(E_1^c) = 1 - \dfrac{243}{400} = .3925$

**b) $E_2$** $\beta$ contains exactly two digit "2"s

Events satisfying criterion can be composed from Disjoint Sets A,B,C

$A:\underbrace{\{"2""2"K\}}_{K \ne 2}$ ; $B\underbrace{\{"2"J"2"\}}_{J \ne 2}$ ; $C\underbrace{\{I"2""2"\}}_{I \ne 2}$ ; $E_2 = A \cup B \cup C$ $\qquad$ No intersections "Add counts"

$P(E_2) = P(A \cup B \cup C) = P(A) + P(B) + P(C)$

$= \dfrac{n_{22K}}{n} + \dfrac{n_{2J2}}{n} + \dfrac{n_{I22}}{n} = \dfrac{9}{400} + \dfrac{9}{400} + \dfrac{3}{400} = \dfrac{21}{400} = .0525$

---

When the order of outcomes is relevant then we *must assign different outcomes* to these orderings. However, if the order is irrelevant *we have a choice to use either* an ordered or unordered sample space, but must be careful to count correctly as illustrated by the marble problem in the top panel.

Set algebra and Venn diagrams facilitate the description of compound events in the bottom panel. A random integer in $\beta \, \varepsilon$ [100, 499] is written as $\beta$ = {I J K} with I = 1,2,3,4 and J,K= 0,1,...,9 yielding 400 possible numbers. In **a)** we are asked to find the probability of the event $E_1$, that number $\beta$ *has at least one* digit that is a "1". We define events A, B, and C to be the sets for which there is a digit "1" in the 1st, 2nd, or 3rd digits respectively; the remaining digits are unspecified and so they may also be "1" or any other number consistent with their position. Obviously, these sets are not mutually exclusive (ME), but they are collectively exhaustive (CE) since they cover all the ways in which there is at least one digit "1" in a number constructed from three digits. The Venn diagram illustrates the situation in which there is a single "1" only in A, or only in B, or only in C, or 2 "1"s in pairs AB, or BC, or CA, or 3 "1"s in the triple ABC. The event $E_1$ is just the union of these three sets, *i.e.*, $E_1$=A∪B∪C and its complement $E_1^c$ = (A∪B∪C)$^c$ is the intersection of the complements $E_1^c$ = A$^c$B$^c$C$^c$. Counting the ways in which there are no "1"s in the number $\beta$ is much simpler than enumerating all the ways in which there is at least one. For digit I there are 3 ways to exclude a "1"; for digits J and K there are 9 ways each so we write down immediately $P(E_1^c)$ = (3∗9∗9)/400= .6075 and hence the desired number of ways for at least one digit "1" is $P(E_1)$ = 1- $P(E_1^c)$=.3925. In **b)** the probability of exactly two "2"s defines $E_2$=A∪B∪C, where A, B, C represent three ways to distribute 2 "2"s in a 3 digit number; they are ME and CE so the probabilities just add to yield $P(E_2)$ = (3+9+9)/400=.0525 .

## 3.3.1 Inclusion / Exclusion Ideas

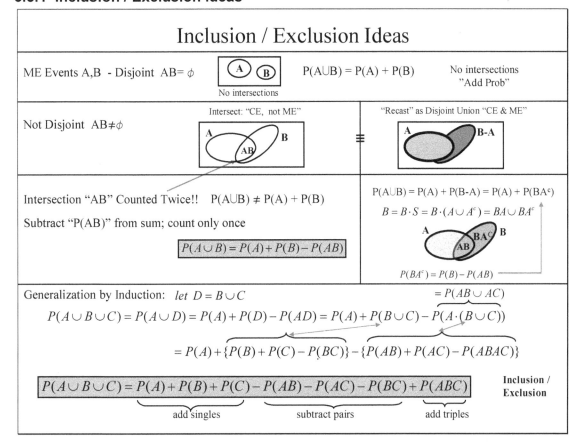

Inclusion / Exclusion Ideas

ME Events A,B - Disjoint $AB = \phi$ — No intersections — $P(A \cup B) = P(A) + P(B)$ — No intersections "Add Prob"

Not Disjoint $AB \neq \phi$ — Intersect: "CE, not ME" — "Recast" as Disjoint Union "CE & ME" — $\equiv$

Intersection "AB" Counted Twice!! $P(A \cup B) \neq P(A) + P(B)$

Subtract "P(AB)" from sum; count only once

$$P(A \cup B) = P(A) + P(B) - P(AB)$$

$$P(A \cup B) = P(A) + P(B-A) = P(A) + P(BA^c)$$

$$B = B \cdot S = B \cdot (A \cup A^c) = BA \cup BA^c$$

$$P(BA^c) = P(B) - P(AB)$$

Generalization by Induction: $let \ D = B \cup C$

$$= P(AB \cup AC)$$

$$P(A \cup B \cup C) = P(A \cup D) = P(A) + P(D) - P(AD) = P(A) + P(B \cup C) - P(A \cdot (B \cup C))$$

$$= P(A) + \{P(B) + P(C) - P(BC)\} - \{P(AB) + P(AC) - P(ABAC)\}$$

$$P(A \cup B \cup C) = P(A) + P(B) + P(C) - P(AB) - P(AC) - P(BC) + P(ABC)$$

Inclusion / Exclusion

add singles — subtract pairs — add triples

It is important to realize that although probabilities are simply numbers that add, the probability of the union of two events P(A∪B) is not equal to the sum of individual probabilities for the two events P(A) + P(B). This is because points in this overlap region AB are counted twice; to correct for this, we need to subtract out "once" the double counted points in the overlap yielding

$$P(A \cup B) = P(A) + P(B) - P(AB).$$

Only in the case of non-intersection $A \cap B = \phi$ does the simple sum of probabilities hold. The generalization for a union of three or more sets alternates inclusion and exclusion. For three sets A,B,C, the probability P(A∪B∪C) adds the $^3C_1$ (=3) singles, subtracts the $^3C_2$ (=3) doubles and adds the $^3C_3$ (=1) triple as shown in the bottom boxed equation. Similarly for four sets A,B,C,D the probability P(A∪B∪C∪D) adds the $^4C_1$ (=4) singles, subtracts the $^4C_2$ (=6) doubles, adds the $^4C_3$ (=4) triples, and subtracts the $^4C_4$ (=1) 4-tuple.

For three sets A,B,C we first define D = B∪C and then apply the two set result directly to A and the composite set D, viz., P(A∪D) = P(A) +P(D) – P(AD); next we substitute the definition of D and expand the resulting expression by re-using the two set result and using the rule for distributing intersections over unions. For n sets $A_1, A_2, ..., A_n$, the general proof proceeds by induction.

## 3.3.2 Probabilities on Union of Intersecting Sets

# Probabilities on Union of Intersecting Sets

Probabilities on union of *non-intersecting sets*
simply add (Axiom#3) $P(A \cup B \cup C) = P(A) + P(B) + P(C)$

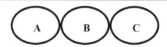

Probabilities on union of *intersecting sets* requires inclusion/exclusion expansion

$$\boxed{P(A \cup B \cup C) = P(A) + P(B) + P(C) - P(AB) - P(AC) - P(BC) + P(ABC)}$$

$\underbrace{\qquad}_{\text{add singles}}$ $\underbrace{\qquad}_{\text{subtract pairs}}$ $\underbrace{\qquad}_{\text{add triples}}$

Re-cast as a union of disjoint sets so that probabilities once again just add

New sets D = A, E = A$^c$B, F = A$^c$B$^c$C are
mutually exclusive as their intersections are null

$$P(A \cup B \cup C) = P(\underbrace{A \cup A^c B \cup A^c B^c C}_{\text{New Sets are Disjoint}}) = P(A) + P(A^c B) + P(A^c B^c C)$$

Yields "clean" separation of events. Similar to Gram-Schmidt
orthogonalization procedure in vector analysis

Instead of using the inclusion/exclusion expansion to evaluate the P(A∪B∪C) on the intersecting sets A, B, C, we can re-cast the problem by defining equivalent disjoint sets D, E, F covering the same region. The probability of the union of these "new" disjoint sets P(D∪E∪F) is simply the sum of the individual probabilities. The construction of the new sets {D, E, F} is straightforward: we first take the "whole set" F=A, then the set B excluding its intersection with A, *viz.*, E=BA$^c$, and finally the set C excluding its intersection with both A and B, *viz.*, F=CA$^c$B$^c$. Note that these new sets {E, F, G} or {A, BA$^c$, CA$^c$B$^c$} are easily shown to be disjoint since they have null intersections with one another, *e.g.*, EF = A (BA$^c$) = B(AA$^c$) = B $\phi$ = $\phi$, *etc.* .

This procedure is somewhat analogous to the Gram-Schmidt orthogonalization procedure in vector analysis.

### 3.3.3 Simple Compound Event Example

# Simple Compound Event Example

**Example:** Student has two books A, B; the probabilities of liking A, B, both A&B are

$$P(A) = .5 \; ; \; P(B) = .4 \; ; \; P(AB) = .3$$

What is the probability the student will like neither A or B?

---

**Solution:**

Neither A or B is represented by event $E = A^c B^c$

$$\underbrace{(A^c B^c)}_{\text{not A \& not B}} = \underbrace{(A \cup B)^c}_{\text{(A or B) not}}$$

$$\underset{DeMorgan}{P(A^c B^c)} = P((A \cup B)^c) = 1 - P(A \cup B)$$

$$= 1 - \{\underbrace{P(A)}_{=.5} + \underbrace{P(B)}_{=.4} - \underbrace{P(AB)}_{=.3}\} = 1 - .6 = .4$$

---

**Set Algebra Note:**

Recall Identity:  $(A \cup B) \cup (A \cup B)^c = S$

Taking Probability of both sides:  $P[(A \cup B) \cup (A \cup B)^c] = P[S]$

$$P[(A \cup B)] + P[(A \cup B)^c] = 1$$

Yields identity used above:  $P[(A \cup B)^c] = 1 - P[(A \cup B)]$

---

This is a simple exercise in applying set algebra to a word problem. The statement that "the student will like neither A or B" is parsed as the intersection of "not-A and not-B" and yields the compound event $A^c B^c$ whose probability we need to calculate. The Venn diagram shows that the set $(A \cup B)$ and its complement $(A \cup B)^c$ constitute the whole space S; it clearly shows that the complement $(A \cup B)^c$ contains no elements of A and no elements of B, *i.e.*, it is the event $A^c B^c$ (which is DeMorgan's Law.)

We only need express the desired probability $P(A^c B^c)$ in terms of the given quantities P(A)=.5, P(B)=.4 and P(AB)=.3. This is easily done using $P(A^c B^c) = P((A \cup B)^c)$ which in turn equals 1- $P(A \cup B)$; expanding the probability of the union gives an expression involving the known quantities in the middle panel and yields the answer 0.4 .

The bottom panel is just a set algebra note stating that the union of an event $(A \cup B)$ and its complement $(A \cup B)^c$ equals the universal set S (also illustrated in the Venn diagram). Taking the probability of both sides of the identity shows the sum of their probabilities is unity, thus allowing us to replace $P((A \cup B)^c)$ with 1- $P(A \cup B)$ which was used in the calculation above.

### 3.3.3.1 Inclusion /Exclusion by Construction-1

# Inclusion/Exclusion by Construction - 1

- Inclusion/Exclusion

N=20 Atomic Events

3 Compound Events: A,B,C

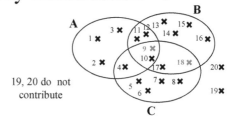

19, 20 do not
contribute

- Want Prob of Union: $P(A \cup B \cup C) = n(A \cup B \cup C)/N = 18/20$

Counting Number of Events in each of 7 regions

$n(A \cup B \cup C) = n(A) + n(B) + n(C) - n(AB) - n(BC) - n(AC) + n(ABC)$

$\phantom{n(A \cup B \cup C) = }\ \ 18 = 8 + 10 + 9 - 4 - 4 - 3 + 2 = 18$

n() IS JUST
NUMBER. IF P(), DIVIDE EACH # BY 20.

- Check that each atomic event only contributes ONCE!!

| ✗ | Pt. #9 in A,B,C | 1 | = | (1 + 1 + 1) | - | (1 + 1 + 1) | + | 1 |
|---|---|---|---|---|---|---|---|---|
| | | | | Singles | | Pairs | | Triples |
| ✗ | Pt. #18 in B,C | 1 | = | (0 + 1 + 1) | - | (0 + 1 + 0) | + | 0 |
| | | | | Singles | | Pairs | | Triples |

The inclusion/exclusion expansion makes sure that each atomic event only contributes once by properly taking into account its appearance in several different events.

**Red Pt#9 is in the triple ABC:** hence Pt#9 appears in each of three singles, in each of the three doubles, and once in the triple. Because of the contribution signs for each "tuple", various contributions cancel one another and as a result Pt#9 contributes only once to the count.

**Red Pt#18 is in the double BC:** hence Pt#18 appears in two singles (B and C), in one double (BC), and of course, it appears in 0 triples. Again, because of the contribution signs for each "tuple", various contributions cancel and Pt#18 contributes only once to the count.

### 3.3.3.2 Inclusion/Exclusion by Construction-2

# Inclusion/Exclusion by Construction - 2

- Atomic Event Contributions

Event #9 occurs in A,B,C

Event #18 occurs in B,C

Each MUST only contribute "1" to LHS "Union"

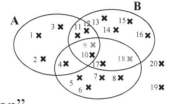

C    19, 20 do not
contribute

LHS                        RHS

$$n(A \cup B \cup C) = n(A) + n(B) + n(C) - n(AB) - n(BC) - n(AC) + n(ABC)$$

#9:     1   =   1    1    1    -1    -1    -1        1

$^3C_0$        $\underbrace{\qquad}_{^3C_1}$    $\underbrace{\qquad}_{-\,^3C_2}$    $\underbrace{\qquad}_{+\,^3C_3}$

Pt#9 is in
3 sets: A,B,C

contributes to:
3 singles, 3 pairs
& 1 triple

#18     1   =   $\cancel{0}$   1    1    $\cancel{0}$   -1    $\cancel{0}$    $\cancel{0}$

$^2C_0$        $\underbrace{\qquad}_{^2C_1}$    $\underbrace{\qquad}_{-\,^2C_2}$    $\underbrace{\qquad}_{+\,^2C_3\,(=0)}$

Pt#18 is in
2 sets: B,C

contributes to:
2 singles, 1 pair
& 0 triples

General: Atomic Event occurs in "m" of $E_{i1}, E_{i2}, \ldots E_{im}$

Recall Identity

$$^mC_0 = 1 = {^mC_1} - {^mC_2} + {^mC_3} - \cdots + (-1)^{m-1}\ {^mC_m}$$

$$0 = (1-1)^m = \sum_{k=0}^{m} {^mC_k}(-1)^k(1)^{m-k}$$

$$0 = {^mC_0} - \left\{ {^mC_1} - {^mC_2} + {^mC_3} - \cdots + (-1)^{m-1}\ {^mC_m} \right\}$$

The contributions of a given "atomic point" (*e.g.*, #9 or #18) to the inclusion/exclusion sum is determined by the number of sets in the union $A \cup B \cup C$ that it occurs in. These contributions can be related to the binomial expansion identity (3) on Slide#2-22

$$0 = (1-1)^m = {^mC_0} - {^mC_1} + {^mC_2} - {^mC_3} + \ldots + (-1)^{k-1}\ {^mC_k} + \ldots + (-1)^{m-1}\ {^mC_m}$$

which, upon moving $^mC_0$ to the LHS, yields

$$^mC_0 = 1 = {^mC_1} - {^mC_2} + {^mC_3} + \ldots + (-1)^{k-1}\ {^mC_k} + \ldots + (-1)^{m-1}\ {^mC_m} \quad \text{for } k = 1,2,3,\ldots,m$$

Applying this formula explicitly to point #9 which occurs in all sets A, B, and C (m=3) yields

$$^3C_0 = 1 = {^3C_1} - {^3C_2} + {^3C_3}$$

For point #18 which occurs in sets B, C (m=2), we have instead

$$^2C_0 = 1 = {^2C_1} - {^2C_2}$$

## 3.3.4 Inclusion/Exclusion - Venn Diagram Application

# Inclusion/Exclusion - Venn Diagram Application

Given following information find how many club members **play at least one sport**  T or S or B

Given:  T=36, S= 28, B=18,
TS=22, TB=12, SB=9, TSB=4

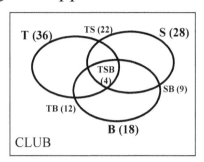

Let N= Total # members (unknown)

Write Probabilities as   $P(T) = \dfrac{36}{N}$ ; $P(S) = \dfrac{28}{N}$;   $P(B) = \dfrac{18}{N}$;  *etc.*

**Method 1:** *Subs into Formula for Union*

$$P(T \cup S \cup B) = P(T) + P(S) + P(B) - P(TS) - P(TB) - P(BS) + P(TBS)$$

PROBABILITY IS A FRACTION

$$= \frac{36}{N} + \frac{28}{N} + \frac{18}{N} - \frac{22}{N} - \frac{12}{N} - \frac{9}{N} + \frac{4}{N}$$

$$= \frac{43}{N}$$

**Thus 43 of "N" Club Members play at least one sport. (N is irrelevant)**

**Method 2:** *Disjoint Union - Graphical*

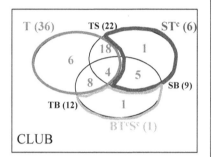

$$T \cup S \cup B = T \cup ST^c \cup BT^cS^c$$

$$P(T \cup S \cup B) = P(T) + P(ST^c) + P(BT^cS^c)$$

$$= \frac{36}{N} + \frac{6}{N} + \frac{1}{N} = \frac{43}{N}$$

This example illustrates the ease with which a Venn diagram can display the probabilities associated with the various intersections of 3 sets T, S, and B.  The problem is to find how many members of a certain club play *at least one* of the three sports tennis (T), squash (S) or baseball (B); the number of club members N is unspecified, but we simply divide the number in each category by N to find the probability.  The event statement "at least one of " means that the we take the union T∪S∪B and since members may play more than one sport, these sets have overlap as shown in the upper Venn diagram.  This diagram is populated with the numbers for each sport and their various intersections given in the problem statement.

**Method 1** applies the inclusion/exclusion expansion formula for P(T∪S∪B) which sums the singles subtracts the pairs and adds the triple to yield the probability of 43/N, so the solution to the problem is 43 members play at least one sport.  Note that the number of elements in each of the 7 distinct regions is easily read off the upper Venn diagram or taken directly from the problem statement.

**Method 2** decomposes the union T∪S∪B into a union of new *disjoint* sets covering the ***same region***, *viz.*, T*∪S*∪B*; thus finding P(T∪S∪B) is equivalent to finding P(T*∪S*∪B*).  But since the starred sets are disjoint, we simply add their individual probabilities P(T*∪S*∪B*) = P(T*) + P(U*) + P(B*).  The number of club members in each set is easily picked off the lower Venn diagram to again yield 43/N.

## 3.3.4.1 Venn Diagram Comparison for 3 and 4 Sets

# Venn Diagram Comparison for 3 and 4 Sets

**Binomial Inclusion/Exclusion Identity #3 (Slide#2-22)**

$$^3C_0 = (^3C_1\ \textit{3-Singles}) - (^3C_2\ \textit{3-Pairs}) + (^3C_3\ \textit{1-Triples})$$

*# Regions = 3 +3+1=7*

$$^4C_0 = (^4C_1\ \textit{4-Singles}) - (^4C_2\ \textit{6-Pairs}) + (^4C_3\ \textit{4-Triples}) - (^4C_4\ \textit{1-Quadruple})$$

*# Regions = 4 +6+4+1=15*

*All 7 Regions represented*

*2 of 15 Regions not represented: $(BD)_O$ & $(AC)_O$*

The $3^{rd}$ and $4^{th}$ order Venn diagrams can be used to visualize the terms in the inclusion/exclusion formulas for P(AUBUC) and P(AUBUCUD) respectively. The union AUBUC can be re-expressed in terms of a ME and CE union of 7 disjoint sets, namely, the **3 singles_only, 3 doubles_only, and 1 triple_only** as indicated by the intersections labeled with subscript "O" in the left panel of the slide. The $3^{rd}$ order Venn diagram has 7 distinct regions that can be uniquely populated as with elements denoted with an "**X**" in the figure; hence the utility of the diagram as a visual aide.

$$AUBUC = \{A_O UB_O UC_O\} \cup \{(AB)_O U(BC)_O U(AC)_O\} \cup \{(ABC)_O$$

On the other hand, the union AUBUCUD can be re-expressed in terms of a ME and CE union of 13 disjoint sets, namely, the **4 singles_only, 4 doubles_only, 4 triples_only, and 1 quadruple_only** as indicated by the intersections labeled with subscript "O" in the figure. In the right panel, two missing pairs are shaded and denoted as missing with symbols BD* and AC* to indicate that the pairs $(BD)_O$ and $(AC)_O$ do not appear in this decomposition as they are *composed of and therefore are not distinct from the others*. More specifically, the decomposition is grouped by singles, pairs, triples, and 4-tuples as

$$AUBUCUD = \{A_O UB_O UC_O UD_O\} \cup \{(AB)_O U(BC)_O U(CD)_O U(AD)_O\}$$
$$\cup \{(ABC)_O U(ABD)_O U(BCD)_O U(ACD)_O\} \cup (ABCD)_O$$

The excluded pairs are actually composed of two triples and the 4-tuple as indicated below

$$BD = (ABD)_O \cup (ABCD)_O \cup (BCD)_O$$
$$AC = (ABC)_O \cup (ABCD)_O \cup (ACD)_O$$

Thus, the Venn diagram for 4 sets is not very useful because it is not capable of distinguishing all six intersection pairs as disjoint regions in the plane. More specifically, we cannot "physically" display an **atomic point** which is a member of only BD or a member of only AC on the Venn diagram of AUBUCUD. We can, of course, display such atomic points (marked with "**X**"s) for the four other pairs AB, BC, CD, and AD since they are disjoint sets. A different arrangement of A,B,C,D will show different pairs, but two pairs will always not appear as disjoint regions. This is because we really need look at the intersection of 4 spheres in 3 dimensions; instead we are just looking at one projection of their intersections. This is in contrast to the Venn diagram for the union of three sets AUBUC, which decomposes into a **ME and CE union** that includes *all 7 regions ( 3 singles_only, 3 doubles_only, and 1 triple_only).*

### 3.3.4.2 Counting Outcomes: Combinations *versus* Permutations

# Counting Outcomes: Combinations *versus* Permutations

**Problem Statement:** *3 draws from urn containing* {6W, 5B}    *Find* P(1W, 2B)

**Method#1: Combinations**

Order of selection **not important**

$$\text{Total Outcomes} = \binom{11}{3} = 165$$

$$\text{Outcomes (1W, 2B)} = \binom{6}{1} \cdot \binom{5}{2} = 60$$

$$P(1W, 2B) = \frac{60}{165} = \frac{4}{11}$$

**Method#2: Permutations**

Order of selection **is important**

If order of selection is counted in outcomes then compute Permutations $^{11}P_3$

$$\text{Total Outcomes} = {}^{11}P_3 = 11 \cdot 10 \cdot 9 = 990$$

Three Orders of Selection for {1W, 2B} as shown in "partial " tree diagram below

$$P(1W, 2B) = \frac{120 + 120 + 120}{990} = \frac{4}{11}$$

**Combinations** are used for counting in an **unordered sample space**; the total #ways of choosing 3 from 11 is $^{11}C_3$ =165. The event $E_1$ = {1W, 2B} results from choosing 1W from 6, $^6C_1$ and then choosing 2B from 5, $^5C_2$; hence the #ways for event $E_1$ to occur is simply the product $^6C_1 \cdot {}^5C_2 = 60$.

On the other hand, doing the same problem in an **ordered sample space** requires us to count all **permutations** of 11 objects taken 3 at a time or a total #ways = $^{11}P_3$ = 990. The event $E_1$ has contributions from all three permutations of {1W, 2B}. The partial tree on the bottom left explicitly shows the paths for the three ordered outcomes, all of which correspond to the same unique state $E_1$ designated by {1W, 2B}. These path contributions {BBW}, {BWB}, and {WBB} are added 120+120+120 to give the total #ways for the event $E_1$ to occur. The probability of the event $E_1$ is therefore 360/990 = 4/11, which is the same value as calculated above using combinations in the unordered sample space.

Thus, although we have three paths with different orderings, we also have three times the number of ordered tree outputs, so the probability obtained by dividing these two numbers cancels the factor of three. In this case, the (partial) tree outputs represent the *three permutations* of the *single group* $E_1$ consisting of 1W and 2B. In general, the full tree will generate all possible nodal sequences (or permutations) and groups are formed from tree outputs by collecting those which have the same "mix" of elements regardless of their order; combinations, in fact, give the number of members in each group based on the mixing of two elements. (Also see Ex. 1 on Slide#3-8).

### 3.3.4.3  Committees from Two groups - Hypergeometric Solution Set

## Committees from Two groups - Hypergeometric Solution Set

Combinations: Take "5" from {6M, 9W}  *Find P(3M,2W)*

$$P(3M,2W) = \frac{\binom{6}{3}\cdot\binom{9}{2}}{\binom{15}{5}} = \frac{240}{1001} = 24\%$$

Binomial Identity #6:

$$\binom{n+m}{r} = \binom{n}{0}\cdot\binom{m}{r} + \binom{n}{1}\cdot\binom{m}{r-1} + \binom{n}{2}\cdot\binom{m}{r-2} + \cdots + \binom{n}{r}\cdot\binom{m}{0}$$

Case: n=6, m=9

$$\binom{15}{5} = \binom{6+9}{5} = \binom{6}{0}\cdot\binom{9}{5} + \binom{6}{1}\cdot\binom{9}{5-1} + \binom{6}{2}\cdot\binom{9}{5-2} + \cdots + \binom{6}{5}\cdot\binom{9}{0} = \sum_{k=0}^{5}\binom{6}{k}\cdot\binom{9}{5-k}$$

**Check**: Sum Prob = "1"

$$\sum_{k=0}^{5} P("k"M,"(5-k)"W) = \sum_{k=0}^{5} \frac{\binom{6}{k}\cdot\binom{9}{5-k}}{\binom{15}{5}} \overset{?}{=} 1$$

$$P(0M,5W) = \frac{\binom{6}{0}\cdot\binom{9}{5}}{\binom{15}{5}} = 4.1\%$$

$$P(1M,4W) = \frac{\binom{6}{1}\cdot\binom{9}{4}}{\binom{15}{5}} = 25\%$$

$$P(2M,3W) = \frac{\binom{6}{2}\cdot\binom{9}{3}}{\binom{15}{5}} = 42\%$$

$$P(3M,2W) = \frac{\binom{6}{3}\cdot\binom{9}{2}}{\binom{15}{5}} = 24\%$$

$$P(4M,1W) = \frac{\binom{6}{4}\cdot\binom{9}{1}}{\binom{15}{5}} = 4.5\%$$

$$P(5M,0W) = \frac{\binom{6}{5}\cdot\binom{9}{0}}{\binom{15}{5}} = 0.1\%$$

$$\Sigma = 100\%$$

The number of ways to choose a committee of 5 consisting of {3M, 2W} from a pool of candidates {6M, 9W} is found directly using combinations. We explore this example further, by writing down the binomial identity (6) (Slide# 2-22) expressing the binomial coefficient $^{n+m}C_r$ as the sum of products $^nC_2 \cdot {}^mC_{r-2}$ representing combinations from two distinct groups with n and m indistinguishable objects respectively. The identity is then written out explicitly for n=6, m=9 corresponding to the pool of candidates {6M, 9W} and it is noted that division of both sides of this equation by $^{15}C_5$ results in the sum of "scaled" product terms that is equal to unity.

This result is actually a well known "probability distribution" which summarizes the probabilities for all possible outcomes to this problem. We explicitly verify all possible "committees of 5 from a pool of 15" on the bottom half of the slide by computing the individual terms P(0M,5W) = 4.1%, P(1M,4W) = 25%, P(2M,3W) = 42%, P(3M,2W) = 24%, P(4M,1W) = 4.5%, P(5M,0W) = 0.1% and showing that they sum to 100% (actually 99.7% with round-off). This distribution can be plotted as a "stick" plot as was done for the binomial distribution; it is, in fact, a common distribution known as the *hypergeometric distribution* and is summarized in the Common PMFs and Properties Table on Slide# 7-6.

## 3.3.5 Urn Problem: k-Draws from Urn with n-Balls –Prob ("Special Ball")

# k-Draws from Urn with n-Balls –Prob( "Special Ball")

**Example**  Urn contains *n* balls, and *one* is **"special"** (red)
Find probability that special ball is chosen in **k-trials**

---

**Method#1**

**(i)** Define Events $\{E_1, E_2, ..., E_n\}$ where, $E_1$ = event that Special Ball drawn in **1st trial**, $E_2$ in **2nd** trial, *etc.*

**(ii)** Events $\{E_1, E_2, ..., E_n\}$ are collectively exhaustive CE since if n trials are made the special ball must appear in one of those trials

$$\therefore \sum_{i=1}^{n} P(E_i) = 1$$

$$P\begin{pmatrix} \text{Spec. ball} \\ \text{in k-trials} \end{pmatrix} = P(E_1 \cup E_2 \cup \cdots \cup E_k)$$

$$= \sum_{i=1}^{k} P(E_i) = \sum_{i=1}^{k} \frac{1}{n}$$

**(iii)** Events $\{E_1, E_2, ..., E_n\}$ are mutually exclusive ME since there is only one special ball, the events $E_1$ & $E_{20}$ cannot both be true

$$= \frac{1}{n} \sum_{i=1}^{k} 1 = \frac{k}{n}$$

**(iv) Equally likely** that each trial yields "special ball"

$$\therefore P(E_1) = P(E_2) = ... = P(E_n) = \frac{1}{n}$$

---

**Method#2**  Direct Counting

$\text{Total Outcomes} = \begin{pmatrix} n \\ k \end{pmatrix}$

$\text{Outcomes that yield special ball} = \begin{pmatrix} 1 \\ 1 \end{pmatrix} \cdot \begin{pmatrix} n-1 \\ k-1 \end{pmatrix}$

$\underbrace{\begin{pmatrix} 1 \\ 1 \end{pmatrix}}_{\substack{\text{Spec} \\ \text{Ball} \\ \text{Drawn} \\ \text{First}}} \cdot \underbrace{\begin{pmatrix} n-1 \\ k-1 \end{pmatrix}}_{\substack{n\text{-}1 \text{ remain} \\ \text{Choose } k\text{-}1}}$

$$P(\text{spec. ball}) = \frac{\begin{pmatrix} 1 \\ 1 \end{pmatrix}\begin{pmatrix} n-1 \\ k-1 \end{pmatrix}}{\begin{pmatrix} n \\ k \end{pmatrix}} = \frac{\dfrac{(n-1)!}{(k-1)!(n-k)!}}{\dfrac{n!}{k!(n-k)!}} = \frac{k}{n}$$

**Note1:** Binomial identity 1) on Slide#2-22 yields → Dividing by $^nC_k$ yields two terms that sum to unity; the second term is precisely the P(spec. ball) obtained above.

$$\begin{pmatrix} n \\ k \end{pmatrix} = \begin{pmatrix} 1+(n-1) \\ k \end{pmatrix} = \underbrace{\begin{pmatrix} 1 \\ 0 \end{pmatrix}\begin{pmatrix} n-1 \\ k \end{pmatrix}}_{\substack{\text{No Spec Ball} \\ R_k^C}} + \underbrace{\begin{pmatrix} 1 \\ 1 \end{pmatrix}\begin{pmatrix} n-1 \\ k-1 \end{pmatrix}}_{\substack{\text{Spec. Ball} \\ R_k}}$$

**Note2:** *ME does not mean Independent Events;* independence not defined yet!  As a preview, consider Tree Representation on next slide

*Two Events are ME and CE*

The probability of drawing a special red ball from an urn containing n balls in k-trials may be computed in a number of different ways.  Arriving at the same result using totally different methods not only gives confidence in its validity, but also gives further insight into the problem and the computation of probability in general.

**Method #1** defines the set of events $\{E_1, E_2, ..., E_n\}$, where $E_i$ means that trial #i yields the red ball; this set of events (together with the null set $\phi$) covers all possible outcomes (CE) resulting from k trials.  Since there is only one red ball, we cannot have both the event $E_2$ (trial #2 yields red) **and** event $E_3$ (trial #3 yields red), and thus these events do not intersect (ME).  Now the probability that the red ball is drawn in trial #1 ($E_1$) is the same as in trial #10 ($E_{10}$), or in fact any other trial; hence the probability for success in any given trial must be 1/n.  It therefore follows that the probability of success in k trials, is simply the probability of the union of all k *disjoint* events $P(E_1 \cup E_2 \cup ... \cup E_k)$ and this evaluates to the sum of "k" equally likely probabilities 1/n, or just k/n.

**Method#2** uses direct counting by combinations (order not important).  The total #outcomes is equal to "n choose k" or $^nC_k$.  The event $R_k$ red ball drawn in k-trials can be computed by assuming the red ball is drawn first $^1C_1$ and the remaining k-1 draws are taken from remaining n-1 balls $^{n-1}C_{k-1}$.  The desired probability $P(R_k)$ is simply the ratio $^1C_1 * {}^{n-1}C_{k-1} / {}^nC_k$ which is again k/n.  Note the binomial identity 2) (Slide#2-22) divided by $^nC_k$ yields a sum of two terms equal to unity.  This represents the "distribution" for the *two* ME and CE events $R_k$ and its complement $R_k^C$ as indicated.

**Method#3** Tree solution on next slide considers the *sequence of events* and involves the concept of conditional probability.

### 3.3.5.1 Tree Solution to Urn Problem

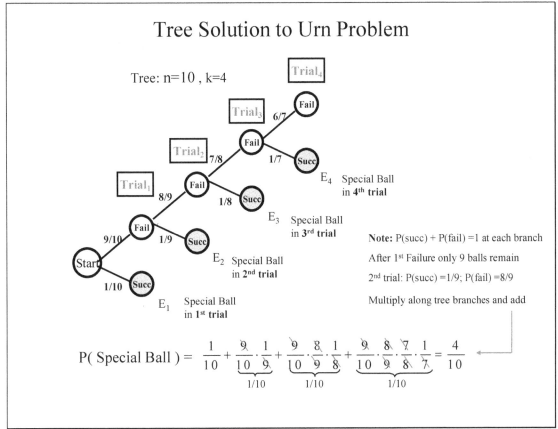

Tree Solution to Urn Problem

Tree: n=10 , k=4

$E_4$ Special Ball in 4th trial

$E_3$ Special Ball in 3rd trial

$E_2$ Special Ball in 2nd trial

$E_1$ Special Ball in 1st trial

Note: P(succ) + P(fail) =1 at each branch

After 1st Failure only 9 balls remain

2nd trial: P(succ) =1/9; P(fail) =8/9

Multiply along tree branches and add

$$P(\text{ Special Ball }) = \frac{1}{10} + \frac{9}{10}\cdot\frac{1}{9} + \frac{9}{10}\cdot\frac{8}{9}\cdot\frac{1}{8} + \frac{9}{10}\cdot\frac{8}{9}\cdot\frac{7}{8}\cdot\frac{1}{7} = \frac{4}{10}$$

1/10     1/10     1/10

The tree for n= 10 balls and k= 4 trials is detailed in this slide along with the results of each trial shown in the "column" below it. The probability of success on the 1st trial is 1/10 (lower branch) and the probability of failure on the 1st trial is 9/10. The probability of success on the 2nd trial can also be shown to be 1/10.

This is easily verified by following "the path to success on the 2nd trial;" the probability of failure on the 1st trial is 9/10, and the probability of success on the 2nd trial (given it failed on 1st trial) is 1/9. The probability of success on the 2nd trial is therefore (9/10)* (1/9) = 1/10.

Continuing in this manner following the paths to success on each trial we see that each trial contributes 1/10 to the success and thus the sum over 4 trial yields 4/10 which is the same result as methods #1 and #2 on the previous slide.

This method sums up the successful outputs of the tree and employs the concept of conditional probability "success given previous failure." Note that the probability of any two tree branches sums to 1 (e.g., 9/10 and 1/10; 8/9 and 1/9, etc.). Thus we simply multiply the probabilities along the legs of the tree diagram to each successful output and then sum up probabilities for all the successes.

The *probability of success given a previous failure* is a "conditional probability" that *always increases* {1/10, 1/9, 1/8,...} because it is computed as one chance in however many remain. However the *probability of success on a given trial* is not conditional; it is computed by multiplying the conditional probability by the *path probability* of arriving at that particular trial. This path probability just compensates for the increasing conditional probability and hence leaves the probability of success on any given trial at exactly 1/10.

## 3.3.6 Re-Characterizing Outcomes by Ignoring a Parameter

# Re-Characterizing Outcomes by Ignoring a Parameter

$n$ Red: $r_1 r_2 r_3 \cdots r_n$    Total # linear arrangements by

$m$ Black: $b_1 b_2 b_3 \cdots b_m$    labeling "color & number" is $(n+m)!$

*Distinguishable Objects: color & number*

Re-characterize outcomes using "colors" only (no numbers)

**Question** : are all color orderings equally likely?

*Distinguishable Objects: color only*

Consider simple case with n = 2 Red and m=2 Black ; then generalize

$$\left( \begin{array}{c} \text{Total \#Outcomes} \\ \text{labeling "color \& number"} \end{array} \right) = 4! = 24 \qquad \begin{array}{c} \textbf{4 distinct} \\ \textbf{objects} \end{array} \quad r_1 \ r_2 \ b_1 \ b_2$$

Now Enumerate these **24 outcomes** by color patterns only to yield **6 outcomes** as follows:

1) Alternate, R first  $r_1 b_1 r_2 b_2$ , $r_1 b_2 r_2 b_1$ , $r_2 b_1 r_1 b_2$ , $r_2 b_2 r_1 b_1$

2) Alternate, B first  $b_1 r_1 b_2 r_2$ , $b_1 r_2 b_2 r_1$ , $b_2 r_1 b_1 r_2$ , $b_2 r_2 b_1 r_1$

**Note 1:** these are all unique outcomes when labeled by *color & number*

3) Paired, R first  $r_1 r_2 b_1 b_2$ , $r_2 r_1 b_1 b_2$ , $r_1 r_2 b_2 b_1$ , $r_2 r_1 b_2 b_1$

4) Paired, B first  $b_1 b_2 r_1 r_2$ , $b_1 b_2 r_2 r_1$ , $b_2 b_1 r_1 r_2$ , $b_2 b_1 r_2 r_1$

**Note 2:** Only each or the 6 rows is unique when labeled *by color*

5) Separate by 2B  $r_1 b_1 b_2 r_2$ , $r_1 b_2 b_1 r_2$ , $r_2 b_1 b_2 r_1$ , $r_2 b_2 b_1 r_1$

6) Separate by 2R  $b_1 r_1 r_2 b_2$ , $b_1 r_2 r_1 b_2$ , $b_2 r_1 r_2 b_1$ , $b_2 r_2 r_1 b_1$

Orderings by "color" are all equally likely with Probability    $P(ordering) = \dfrac{4}{4!} = \dfrac{1}{6}$

This example has a sample space defined by all linear arrangements of objects that are distinguished by two parameters (color and number.) If all arrangements by color and number are equally likely, the question is "are the outcomes still equally likely if we collapse the sample space down one dimension by ignoring (or erasing) the numbers and only considering color arrangements?"

For definiteness, consider the case for n=2 red and m=2 black which yields (4)! = 24 arrangements by number and color and all outcomes are assumed equally likely and have probability 1/24. These 24 distinct arrangements map into 6 arrangements by color as shown on the slide so that arrangement by color alone will also have equally likely outcomes of 1/6.

Consider the outcomes for a fair pair of 6-sided dice designated by the two parameters specifying the face value pairs $(d_1, d_2)$ where $d_1, d_2 = 1,2,3,4,5,6$. The probability for each of these 36 equally likely outcomes is of course 1/36. Re-characterizing these outcomes by ignoring the value $d_2$ yields just six outcomes which all have a probability of 1/6. This should not seem surprising since the two dies take on face values independent of one another and hence ignoring one cannot possibly affect the face value of the other. The equally likely nature remains unchanged for both examples because the two parameters (color, number) or $(d_1, d_2)$ range freely and independently of one another! On the other hand, if we re-characterize the dice outcomes by only reporting their sum $s = d_1 + d_2$, then we would have 11 "sum" outcomes {2,3,4,5,6,7,8,9,10,11,12} that are no longer equally likely since the sum depends upon the value of the second die.

# Numerical Assignment of Probability

## 3.3.7 Bridge Hands

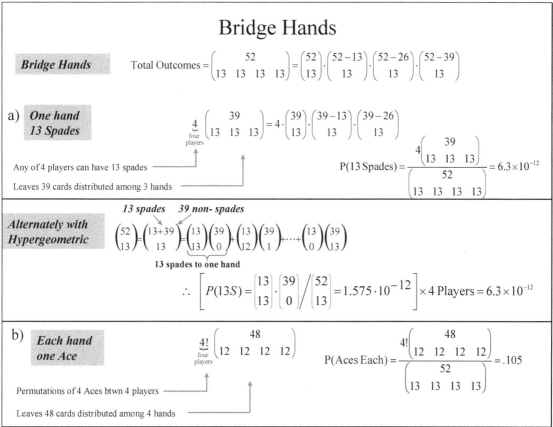

Bridge is a game in which the *entire deck of 52 cards* is dealt randomly among 4 players who play against each other in pairs according to their position North (N), South (S), East (E), and West (W) around a card table. The arrangement of cards within each players hand is unimportant even though each player rearranges the dealt cards by suits and then by number. Because order within each hand is irrelevant the total number of distinct arrangements of cards among the four player positions N, S, E, W is obtained by dividing total number of card permutations 52! by the $(13!)^4$ which is also conveniently expressed as the multinomial coefficient $^{52}C_{13\,13\,13\,13} = 52!\,/(13!)^4$. The same answer results from applying the Hypergeometric binomial expansion (identitity (6) on Slide#2-22) breaking up the 52 cards into 13 spades and 39 non-spades $^{52}C_{13} =^{13+39}C_{13}$, expanding and choosing the appropriate term for one player having 13 spades, *viz.*, $^{13}C_{13}{}^{39}C_0$, then dividing by $^{52}C_{13}$ and multiplying by 4 players.

**Probability that all 13 spades are in 1 hand** is computed by first choosing one hand to have all 13 spades which can be done in $^4C_1$ ways and then distributing the remaining 39 *non-spades* among the other 3 hands, in $^{39}C_{13\,13\,13}$ ways. Hence, the probability is the product of these two numbers divided by the number of ways to *randomly distribute* 52 cards to four hands or $^4C_1\,{}^{39}C_{13\,13\,13}/{}^{52}C_{13\,13\,13\,13}$.

**One Ace in each hand** is computed by first permuting the 4 aces among the 4 hands which is 4! and then distributing the remaining 48 non-Aces among the players $^{48}C_{12\,12\,12\,12}$. Then, dividing by the total #ways, we obtain the probability to be $4!\,{}^{48}C_{12\,12\,12\,12}/{}^{52}C_{13\,13\,13\,13}$. The Hypergeometric expansion can also be used here breaking up the 52 cards into 4 aces and 48 non-aces as follows:

One ace in 1st Hand: $^{52}C_{13} =^{4+48}C_{13} = {}^4C_0{}^{48}C_{13}+{}^4C_1{}^{48}C_{12}+{}^4C_2{}^{48}C_{11}+{}^4C_3{}^{48}C_{10}+{}^4C_4{}^{48}C_9 \rightarrow 4\,{}^{48}C_{12}/{}^{52}C_{13}$

One ace in 2nd Hand: $^{39}C_{13} =^{3+36}C_{13} = {}^3C_0{}^{36}C_{13}+{}^3C_1{}^{36}C_{12}+{}^3C_2{}^{36}C_{11}+{}^3C_3{}^{36}C_{10} \rightarrow 3\,{}^{36}C_{10}/{}^{39}C_{13}$

One ace in 3rd Hand: $^{26}C_{13} =^{2+24}C_{13} = {}^2C_0{}^{24}C_{13}+{}^2C_1{}^{24}C_{12}+{}^2C_2{}^{24}C_{11} \rightarrow 2\,{}^{24}C_{12}/{}^{26}C_{13}$ *etc.*,

Taking the product and rearranging terms we find

$P(\text{1Ace in each Hand}) = 4\,(^{48}C_{12}/{}^{52}C_{13})\,3\,(^{36}C_{10}/{}^{39}C_{13})\,2\,(^{24}C_{12}/{}^{26}C_{13})$

$= 4!\,\{^{48}C_{12}{}^{36}C_{10}{}^{24}C_{12}\}/\{^{52}C_{13}{}^{39}C_{13}{}^{26}C_{13}\} = 4!\,\{^{48}C_{12,12,12,12}/{}^{52}C_{13,13,13,13}\}$

## 3.3.8 Football Roommate Pairings with Restrictions

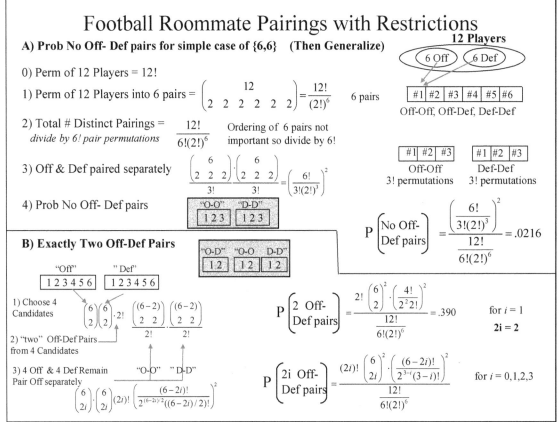

This pairing problem is more subtle than it first appears. A football team consists of 6 defensive players $\{d_1, d_2, d_3, d_4, d_5, d_6\}$ and 6 offensive players $\{o_1, o_2, o_3, o_4, o_5, o_6\}$ and we are asked to find

**A) Probability that a random selection of roommates will have no Off-Def pairs**. The #ways to produce 6 pairs is given by the multinomial coefficient $^{12}C_{222222} = 12!/(2!)^6$

    1) The arrangement of these pairs in rooms is irrelevant so divide by 6!

    2) Thus, the total number of distinct pairings is given by $12!/(2!^6 \, 6!)$

    3d) Pairing among 6 defensive players only (Def-Def) yields $6!/(2!^3 \, 3!)$

    3o) Pairing among 6 offensive players only (Off-Off) also yields $6!/(2!^3 \, 3!)$

    4) Hence, the probability of *no Off-Def pairs* is simply $= [6!/(2!^3 \, 3!)]^2 / [12!/(2!^6 \, 6!)] = .0216$

    Thus, this is very unlikely to happen by accident.

**B) Exactly Two Off-Def Pairs**. First select 2 Off from a "pool" of 6, in $^6C_2$ ways and then 2 Def from a "pool" of 6, in $^6C_2$ ways yielding a total of $(^6C_2)^2$ possible such sets. From each of these sets containing 2 Off and 2 Def players we can choose pairs in just $2 \Rightarrow 2!$ ways and thus the total number of Off-Def pairs is just the product of (pairs/set)*(#sets) $= 2!(^6C_2)^2$. For the remaining 4 Off and 4 Def players we do not want any pairs and this is identical to 3d), 3o) above with 6 replaced by 4, and yields $(^4C_{22}/2!)^2$. Collecting these product factors and dividing by total #outcomes in (2) above gives

$$\text{Prob(exactly 2 Off-Def prs)} = (^6C_2)^2 \, 2! \cdot \{(^4C_{22}/2!)^2 / (^{12}C_{222222}/6!)\} = .39$$

Alternately, using double factorial notation of the next slide (3-23), the brace term may be simply written as $\{(3!!)^2/(11!!)^2\}$ so we have Prob(2 Off-Def prs) $= (^6C_2)^2 \, 2! \cdot \{(3!!)^2/(11!!)^2\} = .39$. Only even numbers 0, 2, 4, 6 Off-Def pairs can occur; 5 Off-Def pairs would leave 1-Off and 1-Def and hence necessarily a $6^{th}$ Off-Def pair. The probability distribution is P[0,2,4,6]=[.02, .39, .52, .07].

## 3.3.8.1 Football Roommate Pairings and Double Factorial

# Football Roommate Pairings and Double Factorial

**Alternate Double Factorial Method:**

**Part a) no Off-Def pairs**

0) Double Factorial Defn:

$$\frac{12!}{6!(2!)^6} = \frac{1}{6!}\frac{12\cdot10\cdot8\cdot6\cdot4\cdot2}{2\cdot2\cdot2\cdot2\cdot2\cdot2}11\cdot9\cdot7\cdot5\cdot3\cdot1 = 11\cdot9\cdot7\cdot5\cdot3\cdot1 = 11!!$$

1) Total # Distinct Pairings case of 4 Off (or 4 Def)

| Pr #1 | Pr #2 |
|-------|-------|
| 1 3 | 1 1 |

# ways  $3\cdot1$

2) Total # Distinct Pairings case of 6 Off (or 6 Def)

| Pr #1 | Pr #2 | Pr #3 |
|-------|-------|-------|
| 1 5 | 1 3 | 1 1 |

# ways  $5\cdot3\cdot1$

Prob stick graph: horizontal axis "Off-Def Pairs" values 0, 2, 4, 6; vertical axis Prob from 0 to .5

$$P\begin{bmatrix}\text{No Off-}\\\text{Def pairs}\end{bmatrix} = \frac{\left(\frac{6!}{3!(2!)^3}\right)\cdot\left(\frac{6!}{3!(2!)^3}\right)}{\frac{12!}{6!(2!)^6}} = \frac{(5!!)^2}{11!!} = \frac{(5\cdot3)^2}{11\cdot9\cdot7\cdot5\cdot3\cdot1} = .0216$$

(Off & Def Paired Separately)

$$P\begin{bmatrix}2\text{ Off-}\\\text{Def pairs}\end{bmatrix} = (^6C_2)^2 2!\frac{(3!!)^{\wedge}2}{11!!} = .390 \qquad P\begin{bmatrix}4\text{ Off-}\\\text{Def pairs}\end{bmatrix} = (^6C_4)^2 4!\frac{(1!!)^{\wedge}2}{11!!} = .519$$

$$P\begin{bmatrix}6\text{ Off-}\\\text{Def pairs}\end{bmatrix} = (^6C_6)^2 6!\frac{(0!!)^{\wedge}2}{11!!} = .0693$$

The relation of the football roommate pairs and the idea of double factorial is established by dividing the multinomial $^{12}C_{2\,2\,2\,2\,2\,2}$ by 6!, since we do not care about the order of the pairings. Note that for 12 objects to be paired, the double factorial result is $(12-1)!! = 11!!$, so for any even number of objects the number of pairings is $(n-1)!!$, where $n-1$ is always an odd number. Once this relationship has been established, the double factorial argument can be applied directly to any pairing situation in which the order of pairings is not important.

For the case of n=12 football players consisting of 6 Off and 6 Def players, the number ways to choose arbitrary pairs is 11!!, while the number of ways to pair among the Def only or Off only are both just 5!! Thus, the probability of having 0 or "No Off-Def pairs" is just the ratio of $(5!!)^2/11!!$ =.0216. Similar calculations for 2, 4, and 6 *mixed* Off-Def pairs make use of the double factorial formula for the remaining Off-Off and Def-Def pairs as shown on the slide.

For example, in the *case of 2 Off-Def pairs,* we first choose 2 from the Off pool of players and 2 from the Def pool of players to yield $(^6C_2)(^6C_2)$; we next find 2! mixed pairings to yield $(^6C_2)^2$ 2!; finally, we use the double factorial expression applied to the remaining 4 Off-Off and 4 Def-Def pairings to yield a factor $(3!!)^2/11!!$; the final result is $(^6C_2)^2$ 2! $(3!!)^2/11!! = .390$.

A stick graph of the probability distribution as a function of the discrete values of 0,2,4,6, Off-Def pairs is also shown on the slide. As is generally the case with such discrete distributions, the extreme cases of 0 and 6 mixed pairs is quite small $\sim$ 10%, while the middle values 2 and 4 mixed pairs together represent the remaining 90%. Somewhat intuitively, this results from the larger number of combinations for the middle values, *i.e.*, compare $^6C_2 = {^6C_4} = 15$ with $^6C_0 = {^6C_6} = 1$.

## 3.3.9 Round Table Husbands and Wives

# Round Table Husbands and Wives

- 10 couples  Round Table No Husband/Wife Sit Together

- Let $E_i$ = *event $i^{th}$ couple sits together*

- Prob none sit together:     $P(None) = 1 - P\left(\bigcup_{i=1}^{10} E_i\right)$

- **3 Couples:** Round Table  Total# Outcomes = $6!/6 = 5!$

| A | B | C |
|---|---|---|

3! Arrangements

3!/3 = 2!
Arrangements

couple 1

| 1 | 2 |  | 3 | 4 | 5 | 6 |

1 Obj        4 Obj

Perm 5 Obj

$\dfrac{(6-1)!}{(6-1)}$

couple 1  couple 2

| 1 | 2 |  | 3 | 4 |  | 5 | 6 |

1 Obj   1 Obj    2 Obj

Perm 4 Obj

$\dfrac{(6-2)!}{(6-2)}$

··· k =1,2,3 "together"

$\Rightarrow \dfrac{(6-k)!}{(6-k)} = (6-k-1)!$

$MW, WM$ each couple $2^k$

$\Rightarrow \boxed{\dfrac{2^k \cdot (6-k-1)!}{5!}}$

$$P(E_1 \cup E_2 \cup E_3) = \underbrace{P(E_1)}_{\frac{2^1 \cdot (6-1-1)!}{5!}} + \underbrace{P(E_2)}_{\frac{2^1 \cdot (6-1-1)!}{5!}} + \underbrace{P(E_3)}_{\frac{2^1 \cdot (6-1-1)!}{5!}} - \underbrace{P(E_1E_2) - P(E_1E_3) - P(E_2E_3)}_{\frac{2^2 \cdot (6-2-1)!}{5!} \quad \frac{2^2 \cdot (6-2-1)!}{5!} \quad \frac{2^2 \cdot (6-2-1)!}{5!}} + \underbrace{P(E_1E_2E_3)}_{\frac{2^3 \cdot (6-3-1)!}{5!}} = .733$$

$k=1$  $=\binom{3}{1}\frac{2^1 \cdot (6-1-1)!}{5!}$        $k=2$  $=\binom{3}{2}\frac{2^2 \cdot (6-2-1)!}{5!}$        $k=3$  $=\binom{3}{3}\frac{2^3 \cdot (6-3-1)!}{5!}$

- **10 Couples:**

$$P(E_1 \cup E_2 \cup \cdots \cup E_{10}) = \binom{10}{1}\frac{2^1 \cdot (20-1-1)!}{19!} - \binom{10}{2}\frac{2^2 \cdot (20-2-1)!}{19!} + \binom{10}{3}\frac{2^3 \cdot (20-3-1)!}{19!} - \cdots - (-1)^9 \binom{10}{10}\frac{2^{10} \cdot (20-10-1)!}{19!} = .6605$$

$\boxed{\binom{n}{k} \cdot \dfrac{2^k \cdot (2n-k-1)!}{(2n-1)!}}$ ;  $k = 1,2,\cdots n$ "together"      **1-P(E)=.3395**

We have exclusively dealt with "linear arrangements" up to this point. If the arrangements are formed in a continuous manner around a table (round or any shaped)  the left and right end of a linear arrangement are connected together. The inset diagram for three objects A, B, C shows that the 3! =6 linear arrangements reduces to 3!/3 = 2  arrangements around a table.  Specifically, in the linear case we have 6 distinct arrangements {{ABC,  BCA, CAB}, {CBA, BAC, ACB}}, while for the round table there are only two distinct arrangements corresponding to one member from each sub-group, e.g., {ABC}, {ACB}.  Other arrangements are just *clockwise* or *counter-clockwise* rotated versions of the two we have chosen.  We can generalize this result to *n* people sitting around a table to be *n!/n* since there are *n* rotations by 360/n degrees that leave the arrangement unchanged.

10 couples are seated around a table and the problem is to find the probability that no husband and wife sit next to one another.  The basic approach is to define the CE set of events {$E_1$, $E_2$, ... , $E_{10}$}, where $E_i$ denotes that the $i^{th}$ couple sits together, compute the $P(E_1 \cup E_2 \cup ... \cup E_{10})$ using the inclusion/exclusion expansion, and then compute the probability for no matches as the complementary probability 1- $P(E_1 \cup E_2 \cup ... \cup E_{10})$.  The simpler case of 3 couples (6 people) is easily generalized if we "track" our calculations; the total #outcomes is 6!/6=5!.  The computation details are best understood in terms of the illustrations, but briefly goes as follows:  The probability $P(E_1)$ that the $1^{st}$ couple sits together is found by considering the couple to be a single object "1-2" (glued together) and then permuting just 5 objects (1 couple + remaining 4 ).  We "track" in **bold** by writing the circular arrangement of "5" as $(6-1)!/(6-1) = (6-1) \cdot ((6-1)-1)!/(6-1) = (6-1)-1)!$; transposing the couple is a new arrangement, so we multiply by $2^1$ and divide by total #outcomes 5! to find $P(E_1)= 2^1(6-1)-1)!/ 5!$. Similarly, $P(E_1E_2) = 2^2(6-2-1)!/ 5!$ and $P(E_1E_2E_3) = 2^3(6-3-1)!/ 5!$.  Substitution yields the result.

## 3.3.10     Win/Loss Runs as Solution to Integer Equations

# Win/Loss Runs as Solution to Integer Equations

$n$ wins "W" and $m$ losses "L": total # of orderings is binomial $\binom{n+m}{n}$

Exemplar : $\underbrace{LL\cdots L}_{\substack{y_1=0,1,2\cdots \\ 0 \text{ is OK!}}} \underbrace{WW\cdots W}_{\substack{x_1=1,2,\cdots \\ \text{positive}}} \underbrace{LL\cdots L}_{\substack{y_2=1,2\cdots \\ \text{positive}}} \underbrace{WW\cdots W}_{\substack{x_2=1,2,\cdots \\ \text{positive}}} \cdots \underbrace{LL\cdots L}_{\substack{y_r=1,2\cdots \\ \text{positive}}} \underbrace{WW\cdots W}_{\substack{x_r=1,2,\cdots \\ \text{positive}}} \underbrace{LL\cdots L}_{\substack{y_{r+1}=0,1,2\cdots \\ 0 \text{ is OK!}}}$

**$r$ runs-of-wins of sizes $x_i$ $i =1,2,\ldots, r$ satisfies the equation**

(1) $x_1 + x_2 + \cdots + x_r = n$     Outcomes $\_x = \binom{n-1}{r-1}$

$y_1$ = Num Losses before 1st Win, $\cdots$

$y_r$ = Num Losses after $r^{th}$ Win (Need it to close off Wins!)

**$r+1$ runs-of-loses of sizes $y_k$ $k =1,2,\ldots, r+1$ satisfies the equation**

(2) $y_1 + y_2 + \cdots + y_{r+1} = m$

but exemplar shows first $y_1$ & last $y_{r+1}$ can both be zero

Transform $\bar{y}_1 = y_1 + 1$ & $\bar{y}_{r+1} = y_{r+1} + 1$

All other $\bar{y}_k = y_k$

$(\bar{y}_1 - 1) + \bar{y}_2 + \cdots + (\bar{y}_{r+1} - 1) = m$

(2') $\bar{y}_1 + \bar{y}_2 + \cdots + \bar{y}_{r+1} = m+2 \Rightarrow$    Outcomes $\_y = \binom{(m+2)-1}{(r+1)-1} = \binom{m+1}{r}$

$$\Pr(r \text{ runs of Wins}) = \frac{\binom{n-1}{r-1} \cdot \binom{m+1}{r}}{\binom{n+m}{n}}$$

**Product of numbers of integer solutions to $x_i$ and $y_i$ run equations divided by total # orderings $^{n+m}C_n$**

This example attempts to make "win" predictions based on the observed pattern of previous wins and losses. Specifically, we compute the probability of having r-runs of wins (of any length) given there are n Wins and m Losses. The total #orderings of n wins and m losses is $^{n+m}C_n$ = (n+m)! / n!m!. A "run of wins" must have losses on either side except at the beginning or end of the pattern. This leads to the exemplar shown on the slide with win runs of length $x_1, x_2, \ldots, x_r$ summing to "n" and loss runs of length $y_1, y_2, \ldots, y_{r+1}$ summing to "m." Thus, the number of run arrangements is the product of the number of integer solutions to the *win runs equations* ($x_i$) and to the *loss runs equations* ($y_i$). The win occupation vector $x_i$ has r non-zero components and the number of win solutions is $^{n-1}C_{r-1}$; the loss occupation vector $y_i$ has r+1 components (sandwiching the wins) and allows the two end values $y_1$ and $y_{r+1}$ to both be zero. This is easily taken care of by the transformations shown on the slide and the number of loss solutions is $^{m+1}C_r$. Therefore, the probability of r-runs (of any length) is obtained by taking the product of these two numbers and dividing by $^{n+m}C_n$ (boxed equation.)

For the case n =8 wins, m=6 losses, we look at two extreme distributions: (i) a single run of 8 wins followed by 6 losses (r = 1) with probability $^7C_{1-1}$ $^{6+1}C_1$ / $^{14}C_8$ = 7/3003=1/429 and (ii) alternation of wins and losses (r=7) with probability $^7C_{7-1}$ $^{6+1}C_7$ / $^{14}C_8$ = 1/429. Thus, since it is very unlikely that 8 wins followed by 6 losses would occur strictly by chance, the fact that it is actually does occur suggests a true "slump", *i.e.*, there is a team problem and a new win is not likely to occur after the last 6 losses. Similarly, for the alternating win/loss pattern, actually observing such an unlikely pattern suggests that it is real and that a loss will follow each win, perhaps indicating that the team is overconfident after a win and is just "resting on its laurels!" The full probability distribution for different r-values is P[r=1, 2, 3, 4, 5, 6, 7] = [.002, .049, .245, .408, .245, .049, .002].

## 3.3.11     Win/Loss Runs – Alternate Method.

# Win/Loss Runs – Alternate Method

**Find Probability for "r" runs-of-wins of any size given:  n=# wins (W)    m=# losses (L)**

**Consider distribution  n=8 wins among  m=6 losses as illustrated**

$$\wedge \, L_1 \, \wedge \, L_2 \, \wedge \, L_3 \, \wedge \, L_4 \, \wedge \, L_5 \, \wedge \, L_6 \, \wedge$$

**Locations $\wedge$ for inserting runs of wins:  6+1 $\rightarrow$ m+1**

$$\begin{pmatrix} \text{choose r} \\ \text{positions for} \\ \text{runs-of-wins} \end{pmatrix} = \; {}^{m+1}C_r$$

**Each of these r locations must be occupied by at least one "W" ; otherwise that location would not represent a "run-of-wins"**

**There remain "n-r" Wins to distribute among these r  bin locations (with bin replacement _i.e._, they could all go in just one of the bins)**

$$\begin{pmatrix} \text{Distribute} \\ \text{n-r wins into} \\ \text{r bins w/repl} \end{pmatrix} = \; {}^r\cancel{C}_{n-r} = {}^{n-1}C_{n-r} = {}^{n-1}C_{r-1}$$

Note: The denominator of Pr[ ] expression is the total number of ways to arrange n wins among m losses. There are (n+m)! permutations and the order of both wins & losses are unimportant.

$$\Pr\begin{pmatrix} \text{r runs of} \\ \text{wins of} \\ \text{any size} \end{pmatrix} = \frac{{}^{n-1}C_{r-1} \; * \; {}^{m+1}C_r}{{}^{m+n}C_n}$$

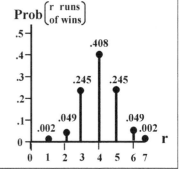

The probability $P_R(r)$ for "r", the number of runs-of-wins of any size may be analyzed using combinations with replacement in a very simple manner as follows. For definiteness, let's consider the case of n=8 wins and m=6 losses and consider the losses $L_1$, through $L_6$ to be the separators between _runs-of-wins_ as illustrated in the upper graphic. The carets "$\wedge$" denote all the "6+1" $\rightarrow$ "m+1"positions into which we may insert the "r" _runs-of-wins_; the number of ways to choose "r" such positions from "m+1"is simply ${}^{m+1}C_r$ as given on the slide. Now each of these r chosen caret positions must be occupied by _at least one_ win W ; otherwise the location would not represent a _"run-of-wins"_. There remain "n-r" wins to place in the now established _runs-of-wins_ positions and these may be done with replacement, _i.e._, all the remaining wins may be placed in a single position or distributed arbitrarily among them. Thus we may consider these as "r" _replaceable bins_ chosen by the remaining "n-r" balls which may be accomplished in slash-${}^rC_{n-r} = {}^{n-1}C_{n-r} = {}^{n-1}C_{r-1}$ ways. Thus the total number of ways to arrange r _runs-of-wins_ among the m losses is just the product of the number of ways to choose the caret positions times the number of ways to distribute the wins in these positions or $({}^{m+1}C_r)({}^{n-1}C_{r-1})$. The total number of ways to distribute the n wins among the m losses is simply the total permutations of (m+n) objects where permutations among the wins and among the losses is irrelevant, _viz._, ${}^{n+m}C_n = (m+n)!/ (m!n!)$. Thus the desired probability for r _runs-of-wins_ is P(r _runs-of-wins_ )= $({}^{m+1}C_r)({}^{n-1}C_{r-1})./{}^{n+m}C_n$. A plot of the probability distribution is shown at the bottom of the slide and it is seen that the extremes of r=1 ,2, 6,7 are quite small with nearly 90% of the probability for runs of size r=3,4,5 (also see Slide#3-23).

## 3.4 Matching Problem

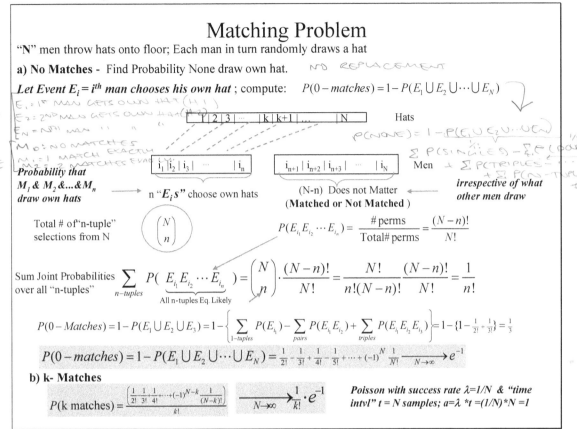

# Matching Problem

"N" men throw hats onto floor; Each man in turn randomly draws a hat

a) No Matches - Find Probability None draw own hat.   NO REPLACEMENT

Let Event $E_i = i^{th}$ man chooses his own hat ; compute:   $P(0-matches) = 1 - P(E_1 \cup E_2 \cup \cdots \cup E_N)$

| 1 | 2 | 3 | ... | k | k+1 | ... | | N |   Hats

$P(NONE) = 1 - P(E_1 \cup E_2 \cup \cdots \cup E_N)$

$\Sigma P(SINGLES) - \Sigma P(DOUBLES)$   1/2!

$+ \Sigma P(TRIPLES) - \cdots$   1/3!

$+ \Sigma P(N-TUPLES)$   $\pm 1/n!$

| $i_1$ | $i_2$ | $i_3$ | ... | $i_n$ |   | $i_{n+1}$ | $i_{n+2}$ | $i_{n+3}$ | ... | $i_N$ |   Men

Probability that
$M_1$ & $M_2$ &...& $M_n$
draw own hats   →   n "$E_i$s" choose own hats

(N-n) Does not Matter   ←
(Matched or Not Matched)

irrespective of what
other men draw

Total # of "n-tuple"
selections from N   $\binom{N}{n}$

$P(E_{i_1} E_{i_2} \cdots E_{i_n}) = \dfrac{\#perms}{Total\#perms} = \dfrac{(N-n)!}{N!}$

Sum Joint Probabilities
over all "n-tuples"   $\displaystyle\sum_{n-tuples} P(\underbrace{E_{i_1} E_{i_2} \cdots E_{i_n}}_{\text{All n-tuples Eq. Likely}}) = \binom{N}{n} \cdot \dfrac{(N-n)!}{N!} = \dfrac{N!}{n!(N-n)!} \dfrac{(N-n)!}{N!} = \dfrac{1}{n!}$

$P(0-Matches) = 1 - P(E_1 \cup E_2 \cup E_3) = 1 - \left\{ \displaystyle\sum_{1-tuples} P(E_{i_1}) - \sum_{pairs} P(E_{i_1} E_{i_2}) + \sum_{triples} P(E_{i_1} E_{i_2} E_{i_3}) \right\} = 1 - \{1 - \tfrac{1}{2!} + \tfrac{1}{3!}\} = \tfrac{1}{3}$

$P(0-matches) = 1 - P(E_1 \cup E_2 \cup \cdots \cup E_N) = \tfrac{1}{2!} - \tfrac{1}{3!} + \tfrac{1}{4!} - \tfrac{1}{5!} + \cdots + (-1)^N \tfrac{1}{N!} \xrightarrow[N\to\infty]{} e^{-1}$

b) k- Matches

$P(k\ matches) = \dfrac{\left( \frac{1}{2!} - \frac{1}{3!} + \frac{1}{4!} + \cdots + (-1)^{N-k} \frac{1}{(N-k)!} \right)}{k!} \xrightarrow[N\to\infty]{} \dfrac{1}{k!} \cdot e^{-1}$

Poisson with success rate $\lambda = 1/N$ & "time
intvl" t = N samples; $a = \lambda *t = (1/N)*N = 1$

---

The man-hat matching problem requires the inclusion/exclusion expansion for a large number of intersecting sets; Venn diagrams are of little use in this case. We shall spend some time on this problem as it is very rich in probability concepts.

The **problem statement** is simple enough: "N men throw their hats onto the floor; each man in turn randomly draws a hat," and we are asked to find the probability of two specific events

a) *That no man draws his own hat.*

b) *That there are exactly k-matches.*

**Key ideas:** Define event $E_i$ = ($i^{th}$ man selects his own hat); take the union of N events $E_1 \cup E_2 \cup ... \cup E_N$ and compute P(no-matches) = 1- $P(E_1 \cup E_2 \cup ... \cup E_N)$.   The **expansion of the $P(E_1 \cup E_2 \cup ... \cup E_N)$** involves addition and subtraction of P(singles), P(pairs), P(triples), *etc.*. (The events $E_i$ are CE, but not ME, so we cannot simply sum up the individual $P(E_i)$ to evaluate this expression.)

A **key part of the proof** is to establish: the sum over singles, P(singles) = 1/(1!); sum over pairs is P(pairs)= 1/(2!); sum over triples is P(triples)=1/(3!); sum over 4-tuples, P(4-tuples) = 1/(4!); ... sum over n-tuples, P(n-tuple) = 1/(n!).   In order to understand these results, consider the case n=3 for the triple $P(E_1 E_2 E_3)$:  There is only 1 way for each of the $1^{st}$ three men to get their own hats; the remaining N-3 men can choose hats in (N-3)! ways, so $P(E_1 E_2 E_3)$ = 1*(N-3)! /N!. But the probability of all $^N C_{N-3}$ triples is the same value $P(E_1 E_2 E_3)$, and hence the sum of triples is just their product, *viz.*,

$\sum [P(E_{i1} E_{i2} E_{i3})] = {}^N C_3 * P(E_1 E_2 E_3) = {}^N C_3 * (N-3)! /N! = N!/[3!(N-3)!] * \{(N-3)! /N!\} = 1/3!$ .

**Limit as N becomes large** approaches a Poisson distribution with success rate for each draw $\lambda$=1/N and data length t =N i.e., parameter a =$\lambda$ t =1.

## 3.4.1 Matching Problem: No Matches

# Matching Problem: No Matches

Use **1) DeMorgan: No Matches** $E_1^c E_2^c \ldots E_N^c = (E_1 \cup E_2 \cup \ldots \cup E_N)^c$ and
**2) Inclusion-Exclusion Theorem:** for Probability of **Union of Non-exclusive events** in which 1 or 2 or 3 or .... or all N men choose own hats: $P(E_{i_1} \cup E_{i_2} \cup \ldots \cup E_{iN})$

$$P(E_1 \cup E_2 \cup \cdots \cup E_N) = \underbrace{\sum P(E_{i_1})}_{1-tuples} - \underbrace{\sum P(E_{i_1} E_{i_2})}_{pairs} + \underbrace{\sum P(E_{i_1} E_{i_2} E_{i_3})}_{triples} - \underbrace{\sum P(E_{i_1} E_{i_2} E_{i_3} E_{i_4})}_{4-tuples} +$$

$$\cdots + (-1)^{n+1} \underbrace{\sum P(E_{i_1} E_{i_2} \cdots E_{i_n})}_{n-tuples} + \cdots + (-1)^{N+1} P(E_{i_1} E_{i_2} \cdots E_{i_N})$$

$$= 1 - \frac{1}{2!} + \frac{1}{3!} - \frac{1}{4!} + \frac{1}{5!} + \cdots + (-1)^{N+1} \frac{1}{N!}$$

$$P(0-matches) = 1 - P(E_1 \cup E_2 \cup \cdots \cup E_N) = \frac{1}{2!} - \frac{1}{3!} + \frac{1}{4!} - \frac{1}{5!} + \cdots + (-1)^{N} \frac{1}{N!} \xrightarrow[N \to \infty]{} e^{-1}$$

Note 1: $\left. e^x \right|_{x=-1} = \left( 1 + x + \frac{x^2}{2!} + \frac{x^3}{3!} + \frac{x^4}{4!} + \cdots \right)_{x=-1} = 1 - 1 + \frac{1}{2!} - \frac{1}{3!} + \frac{1}{4!} - \frac{1}{5!} + \cdots + (-1)^{N} \frac{1}{N!}$

Note 2: $\underbrace{\sum P(E_{i_1} E_{i_2} \cdots E_{i_n})}_{n-tuples} = \frac{1}{n!}$ $\boxed{Case: n = 1 \quad \underbrace{\sum P(E_{i_1})}_{1-tuples} = \frac{1}{1!} = 1}$

Note 3: $\sum_{i=1}^{N} \underbrace{P(E_i)}_{=\frac{1}{N} \atop \text{Equally Likely}} = \sum_{i=1}^{N} \frac{1}{N} = \frac{1}{N} \sum_{i=1}^{N} 1 = N \cdot \frac{1}{N} = 1$

The event "no matches" can be thought of as the intersection of all the complements $E_i^c$ and hence by DeMorgan's laws, the complement of the union of all $E_i$. Thus, using the results for the sums of singles, pairs, *etc.* derived on the last slide (1/k! for an k-tuple), we can immediately write down the probability of the union and obtain the desired result P(no-matches)=1- $P(E_1 \cup E_2 \cup \ldots \cup E_N)$. In the limit of large N the result is shown to be $e^{-1}$.

The fact that the sum (over 1-tuples) =1/1!= 1 can be made intuitive by the following argument. It is equally probable for any of the N men to choose his own hat and hence the probability $P(E_1) = P(E_2) = P(E_3) = \ldots = P(E_N)$ =1/N. Thus summing all singles yields 1. Note that this is somewhat analogous to the urn problem where we found the probability of success (red ball) in the 1st ,10th, or Nth draw was $P(E_1) = P(E_{10}) = P(E_N)$ =1/N; hence for N draws the sum $\sum P(E_i) = N * 1/N = 1$. The probability of success in k draws is k/N, while for N draws it is certain, *i.e.*, "1." Here each man draws but one hat; if allowed to draw k hats the probability of drawing his own hat would be k/N as in the urn problem!

Since the sum of singles = 1, it is clear that such a sum cannot represent the probability of "N-Matches" i.e., all N men get their own hat cannot be 1. It is just as clear that the sum of k-singles cannot be the correct probability for "k-matches." The set of events we want to compute {" 0-matches", "exactly 1- match", ... ,"exactly k-matches",..., "exactly n-matches"} are ME and CE while the set of events {$E_i$} for i=1,2,..,n are CE but not ME. There is, however, a unique connection between the "exactly 0-match" event and the events {$E_i$}, namely, P(0-matches)=1- $P(E_1 \cup E_2 \cup \ldots \cup E_N)$. We shall explore other ways of viewing this problem using Venn diagrams, Tables, and Trees, and discuss the subtle relationship between the sets $E_i$ and "exactly k-matches".

## 3.4.2 Matching Problem: Exactly 'k" matches

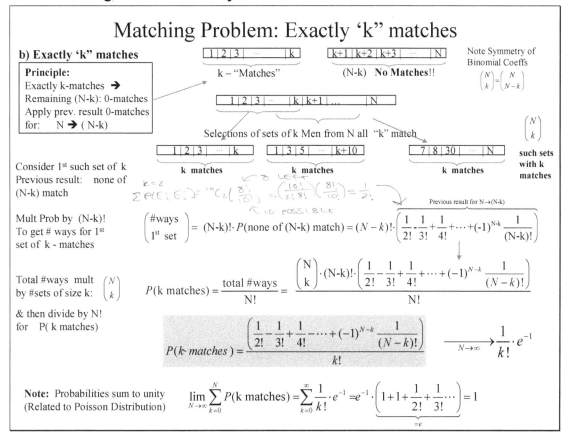

For the event "exactly k-matches", consider that we have chosen k men from N and then matched them with their hats; in order to insure that there are *no more than* k-matches, we must not have any matches in the remaining N-k elements. We have thus separated the "exactly k-matches" result into two distinct sets, namely, one containing exactly k elements with k-matches, and the other containing N-k elements with "0-matches" as shown in the top illustration. The #ways to make such a decomposition is equal to the product of the number of ways we can choose these two sets. The 1st set of "exactly k-matches" from N men can be chosen in $^N C_k$ ways (see lower part of figure), while the 2nd set of "0-matches" from N-k elements is precisely the result already derived for "0-matches" with N-k replacing N. Multiplying these two results together gives the total #ways for "exactly k-matches" and division by the total number of outcomes N! yields the desired probability for "exactly k-matches" given in the shaded boxed equation.

Details of the algebraic manipulations are given on the slide. We also note that in the limit as N→∞ the result is a Poisson distribution with parameter $a = \lambda t = 1$ and the sum of the infinite sequence of terms for k-matches, k =0, 1, 2,... ∞, is unity as it must be.

## 3.4.2.1 Matching Problem: Sum of n-tuples =1/n! for n=1

# Matching Problem: Sum of n-tuples =1/n! for n=1

- Definition of Events:
  - $E_1$ =event 1$^{st}$ man $M_1$ draws own hat $H_1$
  - $E_2$ =event 2$^{nd}$ man $M_2$ draws own hat $H_2$ $\cdots$
  - $\vdots$
  - $E_i$ =event i$^{th}$ man $M_i$ draws own hat $H_i$

$$\sum_{n-tuples} P(E_{i_1} E_{i_2} \cdots E_{i_n}) = \frac{1}{n!}$$

$$Case: n=1 \quad \sum_{1-tuples} P(E_{i_1}) = \frac{1}{1!} = 1$$

- Interpretation of events crucial
- (Recall "3-prisoner paradox")
- Consider case n=1

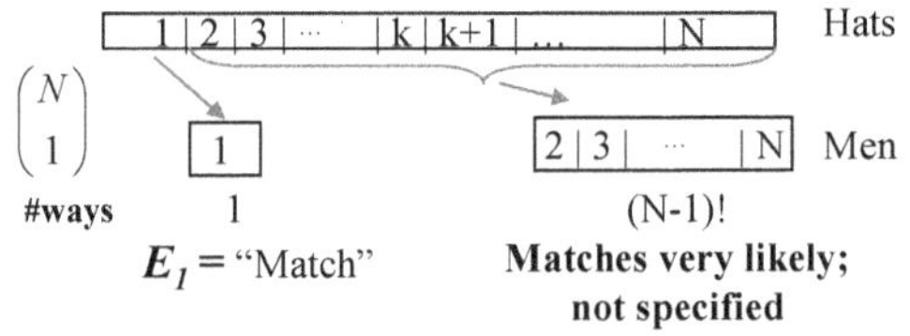

$$\binom{N}{1}$$

#ways

$E_1$ = "Match"

Hats: 1 | 2 | 3 | $\cdots$ | k | k+1 | $\cdots$ | N

Men: 1 | 2 | 3 | $\cdots$ | N

1     (N-1)!

Matches very likely; not specified

$$\sum_{1-tuples} P(E_{i_1}) = \sum_{1-tuples} \frac{(N-1)!}{N!} = \binom{N}{1} \cdot \frac{(N-1)!}{N!} = \frac{N!}{1!(N-1)!} \frac{(N-1)!}{N!} = \frac{1}{1!} = 1$$

- Sum of Equally Likely Events:
  - $P(E_1)=P(M_1$ draws $H_1)=1/N$
  - $P(E_2)= P(M_2$ draws $H_2)=1/N$ $\cdots$
  - $P(E_i)= P(M_i$ draws $H_i)=1/N$

$$\therefore \sum_{i=1}^{N} \underbrace{P(E_i)}_{=\frac{1}{N} \text{ Equally Likely}} = \sum_{i=1}^{N} \frac{1}{N} = \frac{1}{N} \sum_{i=1}^{N} 1 = N \cdot \frac{1}{N} = 1$$

As a reality check on the *sum of n-tuples*, we invoke Principle#0 and compute the simple n=1 case for the sum of singles here and for pairs (n=2) in the next slide. The event $E_1$ denotes that man#1 gets his own hat (hat#1) and similarly $E_2$ states that man#2 does likewise; these events *say nothing about how many matches there are as a result of the men making their selections*. Moreover, the events in the set $\{E_i\}_{i=1,2,\ldots,N}$ are **collectively exhaustive**, but they are **not mutually exclusive** because all outcomes are enumerated but, for example, $E_1$ and $E_2$ may both occur (*i.e.*, they intersect).

For the case n=1 we are interested in the single events $E_i$ only. The *total number of ways* for N men to randomly choose hats equals all the possible permutations of the N hats or N!. As shown in the graphic, there is only one way for man#1 to pick his own hat, and the remaining N-1 men may choose their hats in (N-1)! ways. Thus, the probability for the event $E_1$ is the #ways it can occur 1·(N-1)! divided by N!, *viz.*, $P(E_1) = 1 \cdot (N-1)! / N! = 1/N$. Clearly, similar calculations can be made for $E_2$, $E_3$, *etc.*, so $P(E_i) =1/N$ for all singles and the sum over singles is $\sum P(E_i) = N \cdot (1/N) =1$. **Q.E.D..**

The result also follows from the observation that all events in the set $\{E_i\}_{i=1,2,\ldots,N}$ are *equally likely*, so the probability of all singles $E_i$ must be $P(E_i) =1/N$ and hence the sum over all N events trivially yields 1.

## 3.4.2.2 Matching Problem: Sum of n-tuples =1/n! for n=2

Continuing the reality check, we compute the *sum of n-tuples* for the case n=2 again using the graphic employed in the general case, and also by summing equally likely events and using the concept of conditional probability.

For n=2, we first note that the sum $\sum P(E_j \cdot E_k)$ is evaluated over $^NC_2$ possible pairs ("N take 2"). Since there is no distinction between pairs we only need calculate the probability for one pair, say $P(E_1 \cdot E_7)$, and multiply it by $^NC_2$ in order to evaluate the sum. Proceeding in this manner, we find the #ways to distribute the hats so that man#1 and man#7 get theirs is (N-2)!, and then divide it by N! to obtain $P(E_1 \cdot E_7) = (N-2)! / N! = 1/[N \cdot (N-1)]$ and thus $\sum P(E_j \cdot E_k) = N \cdot (N-1)/2! \cdot 1/[N \cdot (N-1)] = 1/2!$. **Q.E.D.**

The result also follows from the observation that all $^NC_2$ pair events in the set $\{E_i E_j\}_{i \neq j}$ are *equally likely*, so the probability of all pairs $E_i E_j$ must be equal. Following the logic of the tree diagram for the urn problem in Slide#3-19 or using the definition of conditional probability defined in Slide#4-3, we can evaluate the joint probability for the pair as $P(E_1 E_2) = P(E_1) \cdot P(E_2|E_1) = (1/N) \cdot (1/(N-1))$. This can also be understood intuitively by noting that $P(E_1) = 1/N$ for the case n=1; after the event $E_1$ has occurred, there is one less hat (N-1) and man#2 only has one draw, so his "conditional" probability is now $P(E_2|E_1) = (1/(N-1))$ rather than $1/N$ (just as in the tree argument).

Both results agree with the general case and together represent the "Principle#0 steps" we might have taken first before attempting to do the general case. The only issue is that we had to invoke conditional probability (just as in the urn example) before we have formally defined and discussed it. The next several slides explore the matching problem using tables, trees, and Venn diagrams.

### 3.4.3  Matching Problem: Table and Tree Representations (Case: N=3)

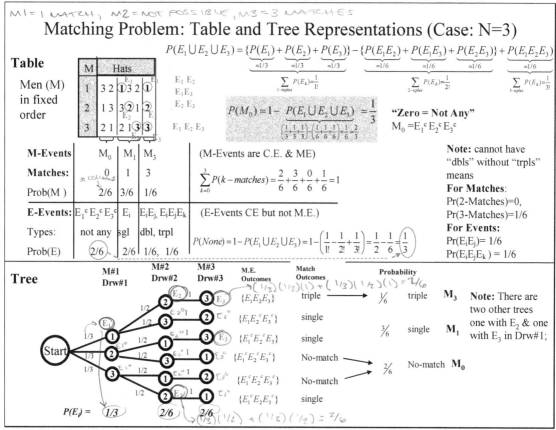

The small table on top displays Man#1, Man#2, Man#3 in fixed order down col#1 and all distributions of hats among the three men down the next 6 columns. We consider two sets of events, (i) M-Events $\{M_0, M_1, M_2, M_3\}$ representing **exactly "0-,1-,2-,3-Matches"** and (ii) E-Events $\{E_1^c E_2^c E_3^c, E_1, E_2, E_3\}$ representing **"no man selects his own hat, M#1 selects Hat#1, M#2 selects Hat#2, M#3 selects Hat#3."** The "matches" $\{M_0, M_1, M_2, M_3\}$ are ME, while the E-Events $\{E_1^c E_2^c E_3^c, E_1, E_2, E_3\}$ are not (intersect); both sets cover all outcomes (CE) and they are connected by the identity that equates "*0-matches*" to "*no man gets his own hat*" or in symbols: $M_0 = E_1^c E_2^c E_3^c$.

The table has circled entries corresponding to events $E_1$, $E_2$, and $E_3$. As we read down the hat columns, we find that cols 1, 2 generate the M-Event $M_0$ and the E-Event $E_1^c E_2^c E_3^c$; cols 3, 4, 5 each generate exactly 1-match $M_1$ and the three single events $E_1$, $E_2$, $E_3$; finally col#6 generates exactly 3-matches $M_3$ (but no $M_2$) and three pairs $E_1E_2$, $E_1E_3$, $E_2E_3$ and one triple $E_1E_2E_3$. Note that we cannot have 2-matches without also having 3-matches, so Prob($M_2$)=0 and P($M_3$)=1/6. The probabilities for the events $E_i$ and their intersections are P[$E_i$]=2/6 (two contributions: 1/6 from 1-match and 1/6 from 3-match), P[$E_iE_j$]=1/6 (contribution: 1/6 from 3-match), P[$E_1E_2E_3$]=1/6 (contribution: 1/6 from 3-match) and finally for no-matches P[$E_1^c E_2^c E_3^c$]=2/6 (contribution from 1st two cols with 0-matches). The bottom half of the slide illustrates how a tree graph expresses the same results; we fix the order of draws M#1, M#2, M#3 just as in the table. Note that the event $E_1$ occupies one node of the tree graph while $E_2$ and $E_3$ occupy 2 nodes each; all have probability 1/3. The match outcomes corresponding to tree outputs labeled by compound events $E_1^c E_2^c E_3^c$, $E_1 E_2^c E_3^c$, *etc.*, are grouped into *no-match, single,* and *triple matches* and yield the correct match probabilities. Note again, that the probability for "2-matches" is zero since it cannot occur without having "3-matches".

## 3.4.3.1 Matching Problem: n =3 Tree/Table Counting

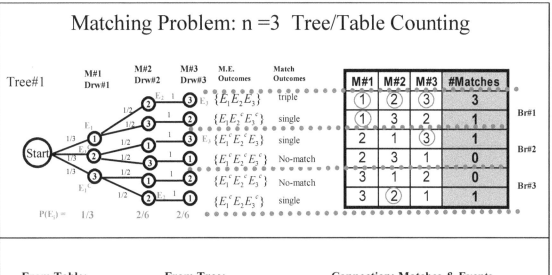

### Matching Problem: n =3 Tree/Table Counting

| | From Table: | From Tree: | Connection: Matches & Events |

**From Table:**

Prob[0-matches]=2/6

Prob[1-matches]=3/6

Prob[2-matches]=0/6=0

Prob[3-matches]=1/6

**From Tree:**

Prob[Sgls]=P[E$_1$]=P[E$_2$]=P[E$_3$]=1/3

Prob[Dbls] = P[E$_1$E$_2$]=(1/3)(1/2)=1/6

Prob[Trpls] = P[E$_1$E$_2$E$_3$]=(1/3)(1/2)=1/6

Alternate Trees Yield: P[E$_1$E$_3$]= P[E$_2$E$_3$]=1/6

**Connection: Matches & Events**

Prob[0-matches]=1-Pr[E$_1$ U E$_2$ U E$_3$]

=1-{Sum[Sngls]-Sum[Dbls]+Sum[Trpls]}

=1-{3(1/3) -3(1/6)+1(1/6)}=2/6

This slide shows the complete the tree and associated table for the Man - Hat problem in which 3 men throw their hats in the center of a room and then randomly select a hat. The drawing order is fixed as Man#1, Man#2, Man#3; the 1$^{st}$ column of tree nodes labeled 1, 2, 3 shows the event $E_1$ (Man#1 draws his own hat) making one contribution of 1/3 at the top node; the two lower nodes are choices for which Man#1 does not draw his own hat, so "taken together" they represent the complementary event $E_1^c$. In the 2$^{nd}$ column of nodes, Man#2 chooses of one of the remaining hats and the event $E_2$ (Man#2 draws his own hat) appears twice making two contributions of 1/6 that sum to give 1/3. Similarly, the 3$^{rd}$ draw by Man#3 yields the event $E_3$ in the two positions and the two contributions again sum to 1/3.

The tree output labeled "M.E. Outcomes" is given by a set of mutually exclusive composite events {$E_1E_2E_3$}, {$E_1E_2^cE_3^c$}, etc., corresponding to the tree paths; the corresponding "match outcomes" are shown in the adjacent column. The table to the right of the tree has columns "M#1, M#2, M#3, and Matches" and its entries can be read directly from the composite outputs as follows: {$E_1E_2E_3$} translates to the first row values "1 2 3 **3**" meaning each man gets his own hat and there are three matches. Similarly, {$E_1E_2^cE_3^c$} translates as "1 3 2 **1**" meaning only Man#1 gets his own hat and just one match. The remaining rows are populated in the same manner with circles around the matches so they are easy to count. The lower panel establishes a link between the "own-hat" event set {$E_i$} and the "match events" {$M_i$} *via* the literal equality "no match = no man gets own hat" or, $P(M_0) = P(E_1^cE_2^cE_3^c)$. Expressing the RHS as 1-P($E_1 \cup E_2 \cup E_3$)] and using the inclusion/exclusion expansion leads to an explicit mathematical relation with the set of singles, $E_i$ , doubles, $E_iE_j$ and the triple $E_1E_2E_3$.

## 3.4.3.2 Matching Problem: Venn for "Own Hat"-Events and for Matches

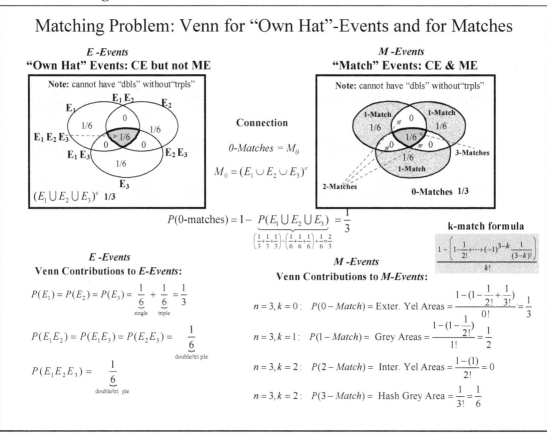

Venn Diagrams for the Man-Hat problem using the "own hat" events $\{E_1, E_2, E_3,$ and composites$\}$ as well as for the match events $\{0$-match, 1-match, 2-match, 3-match$\}$ are displayed here for comparison; all the features and partial contributions previously discussed are in evidence.

The probabilities for singles, doubles, triples can be directly read off the left figure for events $E_i$ and composites; the fact that doubles cannot exist without triples is also obvious from the "0" values in the *parts of the double regions that lie outside the triple region.*

Similarly for matches in the Venn diagram on the right, the 0-,1-,2-,3-match regions and their probabilities are evident. However the M-events are both CE and ME so that a direct sum of their individual probabilities yields 1.

The connection between these two Venn diagrams is found by recognizing that the event $M_0$ signifying "0-matches" is equivalent to the composite E-space event $(E_1 \cup E_2 \cup E_3)^c$. Thus we express this equivalence by the formula

$$P(M_0) = P(0\text{-matches}) = P[(E_1 \cup E_2 \cup E_3)^c] = 1 - P(E_1 \cup E_2 \cup E_3).$$

## 3.4.3.3 Matching Problem: n=4 Tree/Table Counting

# Matching Problem: n=4 Tree/Table Counting

| M#1 | M#2 | M#3 | M#4 | #Matches |
|-----|-----|-----|-----|----------|
| 1 | 2 | 3 | 4 | 4 |
| 1 | 2 | 4 | 3 | 2 |
| 1 | 3 | 2 | 4 | 2 |
| 1 | 3 | 4 | 2 | 1 |
| 1 | 4 | 2 | 3 | 1 |
| 1 | 4 | 3 | 2 | 2 |
| 2 | 1 | 3 | 4 | 2 |
| 2 | 1 | 4 | 3 | 0 |
| 2 | 3 | 1 | 4 | 1 |
| 2 | 3 | 4 | 1 | 0 |
| 2 | 4 | 1 | 3 | 0 |
| 2 | 4 | 3 | 1 | 1 |

Br#1 (top 6 rows)
Br#2 (next 6 rows)

Same #Matches as Br#2 — Br#3
Same #Matches as Br#2 — Br#4

$P[E_1]=1/4$
$P[E_2]=3/12=1/4$
$P[E_3]=6/24=1/4$
$P[E_4]=6/24=1/4$

Extension of the Man - Hat problem to n= 4 where 4 men throw their hats in the center of a room and then randomly select a hat is completely described by fully developing the two upper branches of the tree as shown. The event $E_1$ corresponding to man#1 chooses hat#1 has a unique appearance in this tree and appears no where else, whereas the corresponding events $E_2$, $E_3$, and $E_4$ all appear at several locations in the four branches of the tree as shown. Thus Br#2 and the two branches labeled Br#3 and Br#4 in the diagram have identical characteristics and contribute the same distribution count for the #matches. These three lower branches cannot have 4-matches; only the top branch (Br#1) can.

The probability for 0-,1-,2-,3-, and 4-matches can be read directly off the "partial table" provided we take into account the "missing" branches by adding three times Br#2 to Br#1. Thus we compute (Br#1 + 3* Br#2) to find

P(0-matches) = [0 +3(3)]/24 = 9/24
P(1-matches) = [2 +3(2)]/24 = 8/24
P(2-matches) = [3 +3(1)]/24 = 6/24
P(3-matches) = [0 +3(0)]/24 = 0/24
P(4-matches) = [1 +3(0)]/24 = 1/24

Alternately, we can use the product of conditional probabilities leading to each final state in the tree to obtain
$P(E_1)=1/4$, $P(E_1 E_2)=1/12$, $P(E_1 E_2 E_3)=1/24$, and $P(E_1 E_2 E_3 E_4)=1/24$

Clearly the results for all **singles, all pairs, all triples, and all 4-tuples** are the same. We can use **inclusion-exclusion law** to find $P(E_1 \cup E_2 \cup E_3 \cup E_4)$ in terms of these joint probabilities there results
P(0-matches) =1- $P(E_1 \cup E_2 \cup E_3 \cup E_4)$ =1- {4*$P(E_1)$-6*$P(E_1 E_2)$+4*$P(E_1 E_2 E_3)$-$P(E_1 E_2 E_3 E_4)$}
P(0-matches) = 1-{4(1/4)-6*(1/12)+4*(1/24) -1/24} = 1-15/24 = 9/24  which agrees with the direct table count.

### 3.4.3.4 Matching Problem: Validate "No-Match Outcomes"

## Matching Problem: Validate "No-Match Outcomes"

Validate "No-Match Outcomes" with Tree Diagrams for N=2,3,4

$$P(\text{No-match}) = 1 - P(E_{i_1} \cup E_{i_2} \cup \cdots \cup E_{i_n}) = \frac{1}{2!} - \frac{1}{3!} + \frac{1}{4!} - \frac{1}{5!} + \cdots + (-1)^N \frac{1}{N!}$$

| Case | Man#1 Drw#1 — Man#2 Drw#2 — Man#3 Drw#3 — Man#4 Drw#4 (tree diagram) | "no-match" outcomes | Total # outcomes | Probability "no-match" | General Formula | Table Fixed Order for N |
|---|---|---|---|---|---|---|
| N=2 | S $E_1^c$(2) — $E_2^c$(1) | 1 | 2!=2 | $\dfrac{1}{2!} = \dfrac{1}{2}$ | $\dfrac{1}{2!} = \dfrac{1}{2}$ | Man # / Hat #: 1→2, 2→1. *1 way of 2!* |
| N=3 | S $E_1^c$(2)—$E_2^c$(3)—$E_3^c$(1); (3)—(1)—(2) | 2 | 3!=6 | $\dfrac{2}{3!} = \dfrac{1}{3}$ | $\dfrac{1}{2!} - \dfrac{1}{3!} = \dfrac{1}{3}$ | Man # / Hat #: 1→(2,3), 2→(3,1), 3→(1,2). *2 ways of 3!* |
| N=4 | S (2),(3),(4) branching through $E_1^c$,$E_2^c$,$E_3^c$,$E_4^c$ tree | 9 | 4!=24 | $\dfrac{9}{24}$ | $\dfrac{1}{2!} - \dfrac{1}{3!} + \dfrac{1}{4!} = \dfrac{12-4+1}{24} = \dfrac{9}{24}$ | Man # / Hat #: 1→(2,2,2...), 2→(1,3,4...), 3→(4,4,1...), 4→(3,1,3...). *9 ways of 4!* |

Here is another way to use partial tree diagrams to validate the no-match probability for the N= 2,3,4 matching problem. The partial tree ignores all branches except those that lead to the not-Events $E_1^c$, $E_2^c$, $E_3^c$, $E_4^c$. The intersection of 2, 3, or 4 of the latter events corresponds to "not any" for the corresponding case N= 2,3,4. For N=1 there is a single path to $E_1^c E_2^c$ for N=3 there are two paths leading to $E_1^c E_2^c E_3^c$ and for N=4 there are 9 paths leading to $E_1^c E_2^c E_3^c E_4^c$. Thus counting these tree paths leading to "not any" and dividing by the total #outcomes for each case yields the probability P(no-matches).

The results can be verified directly from the tree, using the general formula, or using the truncated tables on the right. They all agree as they should.

# 4 Probability Types, Interpretations, and Visualizations

# Probability Types, Interpretations, and Visualizations

## 4.1 Conditional Probability

---

## Conditional Probability

• *All probabilities are conditional* because the definition of "everything" must be taken in context. We change context by restricting parameters and their ranges.

• Parameter restriction yields a smaller sample space $\hat{S}$ with fewer outcomes than original sample space S

$$\hat{S} \subset S \quad means \quad P(\hat{S}) < P(S) = 1$$

• *Conditional probabilities must sum to unity* in the reduced sample space $\hat{S}$

• Therefore "renormalize" all probabilities dividing them by $P(\hat{S})$

---

**Example:** Sample Space "S" has 7 *equally likely* points "x"
Event A (ellipse) has 3 points and probability 3/7.

Sample Space "$\hat{S}$" has 3 *equally likely* points (shaded rectangle)

$\hat{S}$ has 2 of its 3 points "in A"; so in $\hat{S}$ the probability of A is 2/3.   **Events:** $\{s_i\}_{i=1,\cdots,7}$

Formally, conditional probability of "A given $\hat{S}$" is calculated by renormalization of joint probability:

$$P(\underset{\substack{read \\ "A\ given\ \hat{S}"}}{A|\hat{S}}) = \frac{P(A \cdot \hat{S})}{P(\hat{S})} = \frac{2/7}{3/7} = 2/3$$

---

The universal set S is the sample space containing points (atomic events) defined by a fixed range of parameters in a multidimensional space. If we reduce the range of one or more of these parameters, then we define a new sample space S-cap that is a subset of the original set S. Clearly, probabilities "conditioned" by the parameter limitations of the sub-space S-cap must be different from those in the original space S. In a very real sense all probabilities are conditional because any "universal set" S can always be considered to be a subset of a higher dimensional or extended parameter set. Thus, we could formally write P(A) as a conditional on the "whole sample space S", *i.e.*, P(A|S), but the conditional is implicit and it rarely written that way. An infinite dimensional parameter space with no limitation on its parameters is clearly the true "Universal Set."

The example on the slide illustrates the computation of the P(A) = P(A|S) and P(A|S-cap). Inspection of the figure shows that the difference between these two probabilities is that event A contains 3 of the 7 points in S so P(A|S) =3/7, and A contains 2 of the 3 points in the subset S-cap, so P(A|S-cap)=2/3. These two conditionals use sample spaces for which P(S) = 7/7 =1 and P(S-cap) = 3/7, and so, mathematically the calculations can be written in the same form as follows:

$$P(A|S) = P(AS) / P(S) = (3/7) / (7/7) = 3/7$$
$$P(A|S\text{-cap}) = P(AS\text{-cap}) / P(S\text{-cap}) = (2/7) / (3/7) = 2/3.$$

It is evident that both calculations "renormalize" or divide by the probability of their respective sample spaces and the only difference is that P(S) =1, while P(S-cap) =3/7. We need never explicitly divide by P(S) because it is assumed we have "covered all possible outcomes" when we declare S as our sample space and so P(S) =1. For the same reason, we never explicitly write P(A|S), just P(A).

## 4.1.1 Conditional Probability: Definition and Properties

# Conditional Probability: Definition and Properties

- Definition of Conditional Probability $\quad P(A\,|\,\hat{S}) \equiv \dfrac{P(A\hat{S})}{P(\hat{S})} \qquad \left(= \dfrac{2}{3}\right)$

- In terms of atomic events $s_i$ we can formally write

$$A = \bigcup_{s_i \in A} s_i \qquad P(A\,|\,\hat{S}) = \frac{P(A\hat{S})}{P(\hat{S})} = \frac{P(\bigcup_{s_i \in A} s_i \hat{S})}{P(\hat{S})} = \frac{\sum_{s_i \in A} P(s_i \hat{S})}{P(\hat{S})} = \frac{(\#\text{pts in } \hat{S} \ \& \ A)}{(\#\text{pts in } \hat{S})}$$

- Note in case $\hat{S} = S$ it reduces to P(A) as it must

- Asymmetry of Conditional Probability

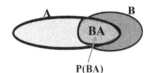

P(BA)

SYMMETRIC

$\Rightarrow P(B\,|\,A) = \dfrac{P(BA)}{P(A)} = \left(\dfrac{\text{fraction}}{\text{BA over A}}\right) = \dfrac{\text{BA}}{\text{A}} \quad \longleftarrow \ \text{Given A}$

*Not Symmetrical!*

$\Rightarrow P(A\,|\,B) = \dfrac{P(BA)}{P(B)} = \left(\dfrac{\text{fraction}}{\text{BA over B}}\right) = \dfrac{\text{BA}}{\text{B}} \quad \longleftarrow \ \text{Given B}$

The formal definition of conditional probability follows directly from the renormalization concept discussed on the previous slide. It is simply the joint probability defined on the intersection of the set A and S-cap, P(AS-cap) divided by the normalizing probability P(S-cap). It can also be written formally in terms of a sum over atomic events as given in the second line of equations.

**Joint probability is symmetric**, *i.e.*, P(A,B) = P(B,A), because the (implied) intersection AB is the same as BA. However, **conditional probability is not symmetric** because the joint probability is divided by the probability of the conditioning set which is P(A) in one case and P(B) in the other. This asymmetry for conditional probability is made especially clear using Venn diagrams and "shape division" where the fixed shape representing P(BA) is divided by that representing P(A) to obtain P(B|A) and by P(B) to obtain P(A|B); the two results are clearly different!

## 4.1.2 Examples: Coin Flips, 4-Sided Dice

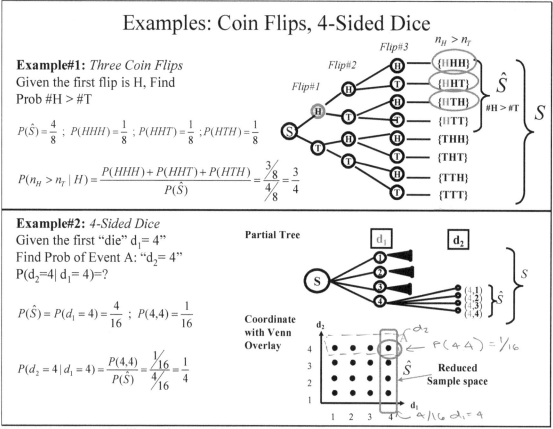

# Examples: Coin Flips, 4-Sided Dice

**Example#1:** *Three Coin Flips*
Given the first flip is H, Find
Prob #H > #T

$$P(\hat{S}) = \frac{4}{8} \; ; \; P(HHH) = \frac{1}{8} \; ; \; P(HHT) = \frac{1}{8} \; ; P(HTH) = \frac{1}{8}$$

$$P(n_H > n_T \mid H) = \frac{P(HHH) + P(HHT) + P(HTH)}{P(\hat{S})} = \frac{3/8}{4/8} = \frac{3}{4}$$

**Example#2:** *4-Sided Dice*
Given the first "die" $d_1 = 4$
Find Prob of Event A: "$d_2 = 4$"
$P(d_2 = 4 \mid d_1 = 4) = ?$

$$P(\hat{S}) = P(d_1 = 4) = \frac{4}{16} \; ; \; P(4,4) = \frac{1}{16}$$

$$P(d_2 = 4 \mid d_1 = 4) = \frac{P(4,4)}{P(\hat{S})} = \frac{1/16}{4/16} = \frac{1}{4}$$

Here are two examples illustrating conditional probability .

**The upper panel** considers a series of three coin flips and the tree shows all possible outcomes for the "3 coin flip" sample space S as an ordered sequence of the nodes. The problem is to find $P(n_H > n_T|H)$, *"the probability that the number of heads (H) exceeds the number of tails (T), given that the first flip resulted in H."* The reduced set of outcomes S-cap only includes the upper branch of the tree emanating from the "head node" (grey circle) of the $1^{st}$ coin flip and specified with an H in the first position: {**H**HH, **H**HT, **H**TH, **H**TT}. The conditional probability in this reduced sample space S-cap is computed either directly by considering only outcomes in S-cap, or by computing the probability relative to outcomes in S (the whole tree) and then renormalizing by the probability of S-cap. The former computation just divides the 3 circled outputs (with **H** in the first position and $n_H > n_T$) by the total of 4 outputs in S-cap to obtain 3/4 directly. The computation in S sums these same three circled outputs assigning them 1/8 probability each and divides by the probability of S-cap which is 4/8, again resulting in $P(n_H > n_T|H) = 3/4$.

**The lower panel** involves a single throw of a pair 4-sided dice and asks for the probability $P(d_2 = 4|d_1 = 4)$, *"that $d_2 = 4$ given that $d_1 = 4$."* Again, the answer is obtained directly from the sub-space S-cap or from the definition of conditional probability in S. The dice sample space is illustrated using a partial tree showing only S-cap as the branches emanating from $d_1 = 4$ and using a coordinate representation with a Venn diagram overlay for the event $(d_1, d_2) = (4,4)$ (green circle) and the subspace S-cap {$d_1 = 4$} (red rectangle). By inspection of the either the tree or the Venn diagram the direct method in S-cap yields a value of 1/4; in S we renormalize the joint probability P(4,4) =1/16 by P(S-cap) =4/16 to again obtain $P(d_2 = 4|d_1 = 4) = 1/4$.

## 4.1.3 Conditional Probability (Constrained Sample Space)

# Conditional Probability (Constrained Sample Space)

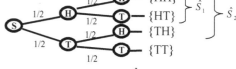

- **Ex. 1) Fair Coin Flipped Twice:**
  **Eq. Likely Outcomes: S = {HH,HT,TH,TT}**
  - A) Find Prob 2 heads "HH" given
    first coin is H: $\hat{S}_1 = \{HH, HT\}$
  
  (***Conditioned on $1^{st}$ coin – upper tree branch***)

  $$P(HH \mid \hat{S}_1) = \frac{P(HH, \hat{S}_1)}{P(\hat{S}_1)} = \frac{1/4}{2/4} = \frac{1}{2}$$

  - B) Find Prob 2 heads given that *at least one*
    coin is H :
  
  (***Conditioned on outcomes***)   $\hat{S}_2 = \{HH, HT, TH\}$

  $$P(HH \mid \hat{S}_2) = \frac{P(HH, \hat{S}_2)}{P(\hat{S}_2)} = \frac{1/4}{3/4} = \frac{1}{3}$$

$\hat{S} = \{5\ S,\ 21\ S^c\}$

- **Ex. 2) Bridge  N-S have 8 spades**
- **Find Prob E has 3 of remaining 5 spades?**

  S → .162   N-S: 8 S, 18 S^c   $P(E_{3S} \mid \hat{S})$ .339   E: 3 S, 10 S^c

  - N-S have 26 cards with 8 spades
  - There remains 26 cards with 5 spades
  - Conditional Sample space $\hat{s}$
    $\hat{S} = \{5\ \text{Spades}, 21\ \text{Non-Spades}\}$

  $$P(\hat{S}) = \frac{\binom{13}{8}\binom{39}{18}}{\binom{52}{26}} = .162 \qquad P(E_{3s} \mid \hat{S}) = \frac{\binom{5}{3}\binom{21}{10}}{\binom{26}{13}} = .339$$

  - (***Conditioned on remaining cards
    after N-S hands are specified***)

  $$P(E_{3S}, \hat{S}) = P(E_{3S} \mid \hat{S}) \cdot P(\hat{S}) = .339 * .162 = .0549$$

**Note:** The conditional probability is "most naturally" calculated from the constrained sample space **directly**. The joint probability is obtained as the product of the conditional probability with that of the sample space.

**Two Methods for Computing Conditional Probability**: i) compute the conditional probability directly in the reduced sample space and ii) compute the joint probability in the original sample space and then divide by the probability of the reduced sample space.

**Two Modes of Conditioning:** Probabilities can be conditioned on i) a previous node or on ii) an event in the full sample space. The coin flip example discussed below illustrates the two methods and two types of conditioning using a simple tree representation. The bridge hand example shows the advantage of using graphical construct to represent the reduced sample space.

**Fair Coin Flipped Twice:** Find the *probability of 2 heads* in two different sample spaces (conditions) **A)** "two heads, given *first coin* is heads" P(HH| $S_1$-cap), or **B)** "two heads, given *at least one coin* is heads" P(HH, $S_2$-cap).

**Given A),** an H on the first flip means that we only need consider the upper tree branch which leads to a reduced sample space $S_1$-cap with only two equally likely outcomes, $S_1$-cap={HH,HT}, so by direct inspection P(HH| $S_1$-cap)=1/2. Using the definition of conditional probability yields the same result P(HH, $S_1$-cap) / P($S_1$-cap) = (1/4) / (2/4)  = 1/2. Note that the conditioning "looks back" at the $1^{st}$ node "H" in this case and not at the tree output.

**Given B),** *at least one* of the coins is H means that we condition on a subset of the 4 outcomes leading to the reduced sample space $S_2$-cap with three equally likely outcomes: $S_2$-cap={HH,HT,TH} each having at least one H; direct inspection yields P(HH| $S_2$-cap) =1/3. Again, using the definition of conditional probability and computing in the full space S we have P(HH| $S_2$-cap) = P(HH,$S_2$-cap) / P($S_2$-cap) =(1/4)/(3/4). In contrast to A), the conditioning here "looks at a subset of outcomes " resulting from both of the primary nodes.

**Bridge Hand with N-S Team Having 8 Spades**: Find the probability "that E has 3 of remaining 5 spades, given N-S has 8 spades" P($E_{3S}$|$NS_{8S}$). The condition space has 5 spades and 21 non-spades, S-cap = {5S, 21S$^c$} and is illustrated in the tree-like graphic. Thus, the conditional probability P($E_{3S}$|$NS_{8S}$) is computed directly from the cards remaining in "S-cap sample space" as P($E_{3S}$|S-cap) = $^5C_3\ ^{21}C_{10} / \ ^{26}C_{13}$ = .339. Similarly, P(S-cap) is computed directly as an event in S given by P(S-cap) = $^{13}C_8\ ^{39}C_{18}/\ ^{52}C_{26}$ =.162. From these two we can find the joint probability from its relation to the conditional probability as P($E_{3S}$,S-cap) = P($E_{3S}$|S-cap) * P(S-cap)=.0549, but there seems to be no easy way to produce the joint probability directly; rather the conditional probability is a more natural calculation and joint probability follows from it.

## 4.1.4 Conditional Probability: Tree Representation

# Conditional Probability: Tree Representation

**Nodes = Events {A,B,D,E,F}**

**Branches = Cond Prob P(B|A) , P(E|DA) , *etc.***

**Compute Joint Probabilities:** follow tree through nodes from start to final state; multiply *a priori* probability by each branch value required to reach the final state

**Example#1:** *Generic Tree*
*Compute Joint Probabilities by following sequential tree*

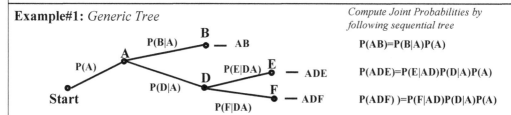

$$P(AB)=P(B|A)P(A)$$

$$P(ADE)=P(E|AD)P(D|A)P(A)$$

$$P(ADF) )=P(F|AD)P(D|A)P(A)$$

**Example#2:** *Draw 2 from {4W, 3R}  Find P(RR)*

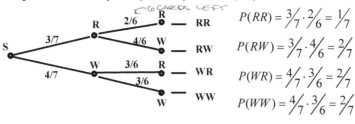

$$P(RR) = \tfrac{3}{7} \cdot \tfrac{2}{6} = \tfrac{1}{7}$$

$$P(RW) = \tfrac{3}{7} \cdot \tfrac{4}{6} = \tfrac{2}{7}$$

$$P(WR) = \tfrac{4}{7} \cdot \tfrac{3}{6} = \tfrac{2}{7}$$

$$P(WW) = \tfrac{4}{7} \cdot \tfrac{3}{6} = \tfrac{2}{7}$$

The concept of conditional probability is easily visualized using a tree. The first branches emanating from the "start" yield nodes representing the *a priori* probabilities and the subsequent nodes are "reached" via branches representing conditional probabilities "given the previous node or nodes." The tree output is designated by the sequence of nodes along the path from start "S" to the output, *e.g.*, {ADE} in the top panel or {RW} in the bottom panel .

**Example#1** in the **middle panel** shows a portion of a generic tree that first leads to event A with branch labeled P(A) (*a priori* probability); the other *a priori* branches are not shown for simplicity, but they sum to unity as do the branch probabilities emanating from any node. Subsequently the tree branches to event B with branch labeled P(B|A) and end state {AB} as well as to event D with branch labeled  P(D|A). There are then two branches from D leading to E and F which are labeled with conditional probabilities P(E|DA) and P(F|DA) and end states {ADE} and {ADF} respectively. Taking products of the probabilities labeling the branches along the path between nodes gives the probability of the *end state node* in terms of the conditional probabilities, e.g., traveling from A to D to F yields P(ADF)=P(F|AD)P(D|A)P(A).

**Example#2** in the **bottom panel** shows a full tree representing all outcomes of 2 draws from a set of white (W) and red (R) marbles from a bowl containing {4W, 3R}. Clearly the 1st draw from 7 balls yields two nodes R or W with *a priori* probabilities of P(R)=3/7 and P(W)=4/7 respectively. The 2nd draw conditioned on the "node R" yields two branches P(R|R)=2/6 and P(W|R)=4/6, while a 2nd draw conditioned on "node W" yields two branches P(R|W)=3/6 and P(W|W)=3/6. The probabilities for each end state are obtained by multiplication of the branch probabilities along the path to that node. Notice that probabilities conditioned on "R" sum to 1, as do those conditioned on "W"; as do the two *a prioris* (which are "implicitly conditioned on S," the entire sample space.)

## 4.1.5 Probability of Winning in the "Game of Craps"

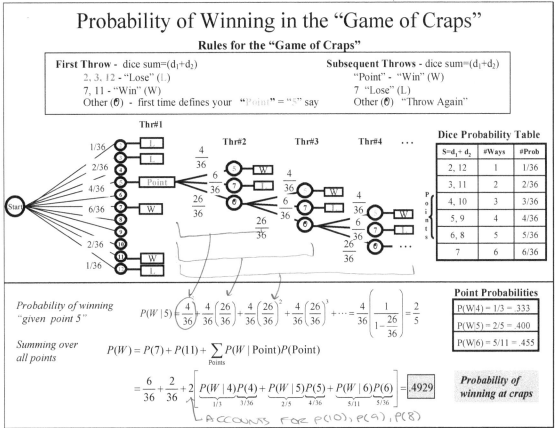

# Probability of Winning in the "Game of Craps"

### Rules for the "Game of Craps"

**First Throw** - dice sum=$(d_1+d_2)$
    2, 3, 12 - "Lose" (L)
    7, 11 - "Win" (W)
    Other ($\theta$) - first time defines your "Point" = "5" say

**Subsequent Throws** - dice sum=$(d_1+d_2)$
    "Point" - "Win" (W)
    7 "Lose" (L)
    Other ($\theta$) "Throw Again"

**Dice Probability Table**

| $S=d_1+d_2$ | #Ways | #Prob |
|---|---|---|
| 2, 12 | 1 | 1/36 |
| 3, 11 | 2 | 2/36 |
| 4, 10 | 3 | 3/36 |
| 5, 9 | 4 | 4/36 |
| 6, 8 | 5 | 5/36 |
| 7 | 6 | 6/36 |

*Probability of winning "given point 5"*

$$P(W\mid 5)=\left(\frac{4}{36}\right)+\frac{4}{36}\left(\frac{26}{36}\right)+\frac{4}{36}\left(\frac{26}{36}\right)^2+\frac{4}{36}\left(\frac{26}{36}\right)^3+\cdots=\frac{4}{36}\left(\frac{1}{1-\frac{26}{36}}\right)=\frac{2}{5}$$

**Point Probabilities**

$P(W|4) = 1/3 = .333$
$P(W|5) = 2/5 = .400$
$P(W|6) = 5/11 = .455$

*Summing over all points*

$$P(W) = P(7) + P(11) + \sum_{\text{Points}} P(W\mid \text{Point})P(\text{Point})$$

$$= \frac{6}{36} + \frac{2}{36} + 2\Big[\underbrace{P(W\mid 4)P(4)}_{1/3 \quad 3/36} + \underbrace{P(W\mid 5)P(5)}_{2/5 \quad 4/36} + \underbrace{P(W\mid 6)P(6)}_{5/11 \quad 5/36}\Big] = \boxed{.4929}$$

*Probability of winning at craps*

ACCOUNTS FOR P(10), P(9), P(8)

Here we use a tree to compute the probability of winning the game of craps that we previously described; for convenience we repeat the rules for the first and subsequent throws in the box at top. Since there are 36 equally likely outcomes, and there is only one way to obtain a sum of 2 or 12, their probability is 1/36; for the dice sums 3 or 11 there are 2 ways so their probability is 2/36; the remaining dice sum probabilities can be read directly off the sum axis coordinate representation on Slide#2-7 and are displayed here in the table to the right of the tree.

The tree shows all the first throw nodes, labels their outcomes, and then places the probabilities along the branches emanating from the "point 5 node" only. The probability for the three outcomes W("5"), L ("7"), "O" (Other: not "5 or 7") can be read off the table as P(5)= 4/36, P(7)=6/36, P(O|5)=1-(4+6)/36=26/36. Note that the probability for the "O" branch is written as conditional P(O|5) since it is different for the other points; however, it does not depend upon previous throws, just on the established point.

Using this tree, we compute the probability of a winning, "given the point is 5" P(W|5), by summing all paths that lead to a win (W) on this "infinite tree." Thus the 2nd throw yields W with probability 4/36 and the 3rd throw yields W with probability P(O|5)P(5) = (26/36)(4/36), and the 4th throw yields W with probability P(O|O,5)P(O|5)P(5) = $(26/36)^2$ (4/36), .... This gives an infinite geometric series $(1+x+x^2+\cdots)$ = 1/(1-x) with x=26/36 and yields P(W|5)=(4/36)*1/(1-26/36)=2/5. Note that P(O|5) = P(O|O,5) = P(O|O,O,5) = $\cdots$ = 26/36 because the results of one trial do not depend upon previous trials.

Now the probability of winning the game itself requires that we sum over all throws and all points. Thus on the 1st throw we sum 6/36 + 2/36 ("7" or "11") and add to this the probabilities of winning on subsequent throws for each possible "point." (Note the factor of "2" in the sum results from taking points in pairs as given in the table.) The infinite sum for each of the other points is obtained in a manner similar to that for "5" and the final result is shown to be .4929, *i.e.*, a 49.3% chance of winning the game of craps! Also note the small table at the bottom which shows the conditionals giving the "best points" for winning.

## 4.1.6 Blood Test & Legal Evidence Interpretation

# Blood Test & Legal Evidence Interpretation

**Blood Test:** Find Probability that if a Person tests positive he has the disease $\quad P(D|T^+)$

*Measurement Statistics:*

| | | |
|---|---|---|
| 98% Pos Test Result if Disease: | $P(T^+|D)=.98$ | *Sensitivity* |
| 3% Pos Test Result No Disease: | $P(T^+|D^c)=.03$ | *Specificity =1-.03=.97* |
| 0.5 % Population has Disease: | $P(D)=.005$ | *a priori % population* |

$$P(D|T^+)=\frac{P(DT^+)}{P(T^+)}=\frac{P(DT^+)}{P(T^+D)+P(T^+D^c)}=\frac{(.98)(.005)}{(.98)(.005)+(.03)(.995)}=.141$$

**Legal Guilt:** Find Probability Defendant is Guilty (G) given the Evidence (E)

$$P(E|G)=.98 \qquad P(E|G^c)=.05 \qquad P(G)=.02 \qquad \textit{200 known thieves in town of 10,000}$$

$$P(G|E)=\frac{P(E|G)P(G)}{P(E|G)P(G)+P(E|G^C)P(G^C)}=\frac{(.98)(.02)}{(.98)(.02)+(.05)(.98)}=.286$$

For a rare disease with probability of occurrence P(D)=.005, a randomly chosen individual has a 0.5% chance of having it, but this probability will change as the result of a good blood test. To be useful, a blood test must be give highly accurate results for two situations, namely (i) *test sensitivity:* a positive test result $T^+$ on a diseased person D, *e.g.*, $P(T^+|D) = .98$, and (ii) *test specificity*: a negative test result $T^-$ on a disease-free person $D^c$, *e.g.*, $P(T^-|D^c)=.97$. The corresponding complementary test results are accordingly (ic) *missed detection*: a negative test result $T^-$ for D, $P(T^-|D)=1-P(T^+|D)=.02$ and (iic) *false positive*: a positive test result $T^+$ for $D^c$, $P(T^+|D^c)=1-P(T^-|D^c)=.03$. Thus the clinical test has two key measurement statistics $P(T^+|D)$ and $P(T^-|D^c)$ that characterize *correct measurements* on two different populations D and $D^c$; they state respectively "how often a diseased person tests positive" and "how often a disease-free person tests negative". The complementary probabilities for these two populations $P(T^-|D)$ and $P(T^+|D^c)$ characterize *incorrect measurements*; they state respectively "how often a diseased person tests negative" (missed detection or *false negative*) and "how often a disease-free person tests positive" (*false positive*) .

The test results can be represented by a tree with two primary nodes D and $D^c$ having, respectively, branch probabilities .005 and .995, representing the proportion of the population with and without the disease. The two branches emanating from node D give the test results as .98 *true positive* and .02 *false negative* (missed detection). Similarly, the two branches emanating from node $D^c$ give the complementary probabilities .03 *false positive* and .97 *true negative*. We have labeled the tree branches with these four conditional probabilities numerically and symbolically, and have represented the four output states by their node sequence as {$DT^+$, $DT^-$, $D^cT^+$, $D^cT^-$}. We also show the sample space of positive tests $T^+$ only. Recall that without tests, we can only say that the probability of having the disease is very small .005. We would like to know how this changes as the result of a positive test; that is, we seek the " $P(D|T^+)$, the probability that a person actually has the disease *given a positive test*. The tree allows us to compute this by only considering the two end states that contain a positive test $T^+$, namely $DT^+$ and $D^cT^+$. In this reduced sample space (dashed shape on tree output), the conditional probability is the probability ratio of positive test emanating from D divided by the two emanating from D and $D^c$ or $P(DT^+) /[ P(D^cT^+)+P(DT^+)] = .141$. This result states that "only in 14.1% of the cases, does a positive test mean the person actually has the disease." This also emphasizes that *conditional probabilities are not symmetric* because $P(T^+|D)=.98$ whereas it's inverse is $P(D|T^+) = .141$ . Lower panel shows similar results for evidence E in determining legal guilt G.

## 4.1.7 Binary Communication Channel

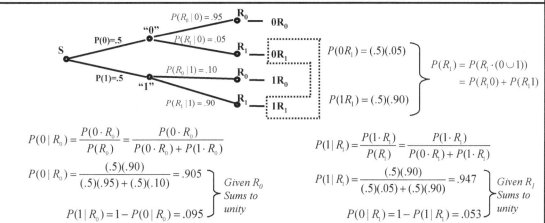

# Binary Communication Channel

**Binary Comm Channel:** Find Probability "1" was sent if receiver detects $R_1$:

Channel Characteristics:

| Measurement Statistics: | | A priori Prob: | Sent | Channel | Rcvd |
|---|---|---|---|---|---|

$P(R_0|0) = .95 \qquad P(R_1|1) = .90 \qquad P(0)=.5$

$P(R_1|0) = .05 \qquad P(R_0|1) = .10 \qquad P(1)=.5$

$\sum_{i=0,1} P(R_i|0) = 1 \qquad \sum_{i=0,1} P(R_i|1) = 1 \qquad P(0)+P(1) = 1$

**Note:** Sum is over 1st variable conditioned on *same* 2nd variable

$$P(0|R_0) = \frac{P(0 \cdot R_0)}{P(R_0)} = \frac{P(0 \cdot R_0)}{P(0 \cdot R_0)+P(1 \cdot R_0)}$$

$$P(0|R_0) = \frac{(.5)(.90)}{(.5)(.95)+(.5)(.10)} = .905 \left.\right\} \text{Given } R_0$$

$$P(1|R_0) = 1 - P(0|R_0) = .095 \left.\right\} \substack{\text{Sums to}\\ \text{unity}}$$

$$P(1|R_1) = \frac{P(1 \cdot R_1)}{P(R_1)} = \frac{P(1 \cdot R_1)}{P(0 \cdot R_1)+P(1 \cdot R_1)}$$

$$P(1|R_1) = \frac{(.5)(.90)}{(.5)(.05)+(.5)(.90)} = .947 \left.\right\} \text{Given } R_1$$

$$P(0|R_1) = 1 - P(1|R_1) = .053 \left.\right\} \substack{\text{Sums to}\\ \text{unity}}$$

The binary communication channel is an application that we will revisit many times. Consider sending a stream of digital "0"s and empirically finding the number of "0"s detected; the resulting frequency of occurrence is a property of the receiver detector circuit. This property is expressed by the "detection statistic," which is defined to be the conditional probability $P(R_0|0) =.95$ (say) that a "0" is received given a "0" was actually sent. Similarly sending a stream of digital "1"s and empirically finding the number of "1"s detected leads to the 2nd detection statistic $P(R_1|1)=.90$ (say) . Notice that the two detection statistics are conditioned on *different signals* "0" or "1" and they each have false detections given by their complements as shown in the upper part of the slide both in the table and the diagram of the "communication channel".

In the lower panel of the slide, we display a labeled tree diagram and compute the inverse probability $P(1|R_1)=.947$ which gives the increased probability of a "1" given the measurement $R_1$ . This result is obtained by computing the probability ratio for tree output nodes containing $R_1$ as in the previous slide, or equivalently by using the conditional probability definition $P(1|R_1) = P(1,R_1)/ P(R_1)$ and recognizing that $P(R_1)$ is precisely the sum $P(0,R_1)+P(1,R_1)$ of path probabilities for the "$R_1$" nodes. The inverse probability for a "0" is found in a similar manner to be

$$P(0|R_0)= P(0\ R_0) / P(R_0) = (.5)(.95) / [(.5)(.95) + (.5)(.10) ] = .905,$$

which is slightly lower than detecting a "1" because of the channel *asymmetry [i.e.,* $P(R0|0)=.95$ *versus* $P(R1|1)=.90$ ]. Note that this example is completely analogous to the blood test example on the previous slide, except that for that case $P(D)$ was very small compared to $P(D^c)$, whereas here the "0"s and "1"s have equal *a priori* probabilities $P(0)=P(1) = .5$ .

## 4.1.7.1 Visualization of Joint, Conditional, and Total Probability

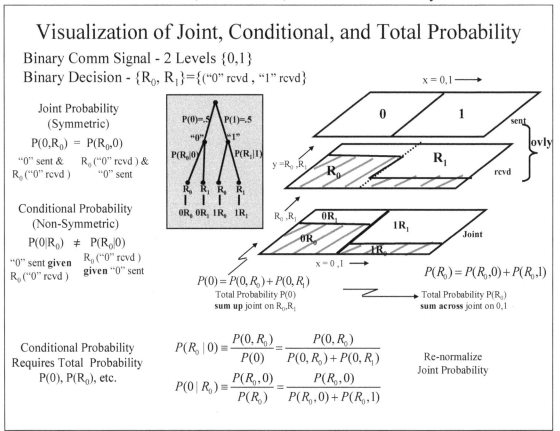

In the previous two examples, we tacitly used the concept of total probability, which we now formalize. Another way to visualize the communication channel is in terms of an overlay of a *Signal Plane* divided (equally) into "0"s and "1"s and a *Detection Plane* which characterizes how the "0"s and "1"s are detected and is structured as shown. The *Outcome Plane* results from overlaying these two planes and yields four distinct regions whose areas represent probabilities of the four product (joint) states {$0R_0$, $0R_1$, $1R_0$, $1R_1$}. These planes correspond precisely to the tree shown in the shaded box to left. The *a priori* probabilities P(0)=P(1)= 0.5 partition the *Signal Plane*, the conditional probabilities $P(R_0|0)$, $P(R_1|0)$, $P(R_0|1)$, $P(R_1|1)$ partition the *Detection Plane*, and the joint probabilities $P(R_0,0)$, $P(R_1,0)$, $P(R_0.1)$, $P(R_1,1)$ partition the *Outcome Plane*.

In this representation the total probability of a "0", P(0) can be thought of as decomposed into two parts summed *into-the-page* along the "0"-half of the bottom plane as indicated by the break-arrow P(0) = P(0,$R_0$) + P(0,$R_1$). Summing on the "1"-half of the bottom plane yields P(1) = P(1,$R_0$) + P(1,$R_1$). Similarly, the total probability P($R_0$) can be thought of as decomposed into two parts summed *horizontally* over the "$R_0$"-portion of the bottom plane P($R_0$) = P($R_0$,0) + P($R_0$,1); we also have P($R_1$) = P($R_1$,0) + P($R_1$,1).

Thus, the intersection of the *signal and detection planes* actually segments the bottom plane into 4 distinct regions labeled by their joint events and the various total probabilities are seen as sums along one of the axes of the horizontal plane. This concept of the decomposition of total probability can be generalized to signals with more complex "n-ary" structure.

Finally, note that conditional probability is defined as the ratio of joint to total probability by the equation $P(1|R_1) \equiv P(R_1,1)/ P(R_1)$, where the total probability is the "sum" $P(R_1) = P(R_1,0) + P(R_1,1)$. Hence, the general concept of conditional probability as the path probability ratio "(desired path)/(sum of paths)" is validated upon substitution of the P($R_1$) "sum" into the conditional probability to yield: $P(1|R_1) = P(R_1,1)/ [ P(R_1,0) + P(R_1,1)]$.

### 4.1.7.2 Joint, Conditional and Total Probability

# Joint, Conditional and Total Probability

**Generalize 3 signal levels**

Tri-nary Comm Signal - 3 Levels $\{-1,0,1\}$   Tri-nary Decision - $\{R_{-1}, R_0, R_1\}$

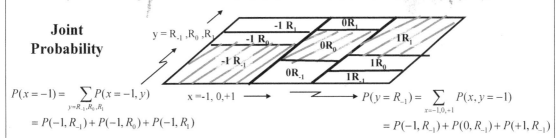

**Joint Probability**   $y = R_{-1}, R_0, R_1$

$P(x=-1) = \sum_{y=R_{-1},R_0,R_1} P(x=-1,y)$   $x=-1, 0, +1 \longrightarrow$   $P(y=R_{-1}) = \sum_{x=-1,0,+1} P(x, y=-1)$

$= P(-1,R_{-1}) + P(-1,R_0) + P(-1,R_1)$   $= P(-1,R_{-1}) + P(0,R_{-1}) + P(+1,R_{-1})$

**Formal Proof of Total Probability Formula for Binary Signal**

Identity   $S = 0 \cup 0^c = 0 \cup 1$

$\Rightarrow R_1 = R_1 \cdot S = R_1 \cdot 0 + R_1 \cdot 1$   $P(R_1) = P(R_1 \cdot (0 \cup 1)) = P(R_1 0) + P(R_1 1)$

Similarly,

Identity   $S = R_0 \cup R_0^c = R_0 \cup R_1$

$\Rightarrow 1 = 1 \cdot S = 1 \cdot R_1 + 1 \cdot R_0$   $P(1) = P(1 \cdot (R_0 \cup R_1)) = P(1R_0) + P(1R_1)$

**Principle of Total Probability**   $P(x) = \sum_y P(x,y) = \sum_y P(x \mid y) P(y)$

The extension of the concepts of joint, conditional, and total probability to a communication signal with *three* (or more levels) is straightforward. The *Signal Plane* is now divided into three distinct levels as "-1"s, "0"s, and "+1"s, and the *Detection Plane* is again a patch-work characterizing how the each signal is detected (not shown). The resulting *Outcome Plane* with six distinct overlay areas that correspond to probabilities of the joint states $\{-1R_{-1}, -1R_0, -1R_1, 0R_{-1}, 0R_0, 0R_1, 1R_{-1}, 1R_0, 1R_1\}$ is illustrated in the figure. Total probabilities for "-1", "0", "1" as well as for "$R_{-1}$", "$R_0$", "$R_1$" are obtained by summing the blocks by y-values y= $R_{-1}$, $R_0$, $R_1$ or by x-values x = -1, 0, 1 across or into the page respectively to yield:

| | | |
|---|---|---|
| P(-1) = P(-1,R₋₁) + P(-1,R₀) + P(-1,R₁) | ; | P(R₋₁) = P(R₋₁,-1) + P(R₋₁,0) + P(R₋₁,1) |

$P(-1) = P(-1,R_{-1}) + P(-1,R_0) + P(-1,R_1)$   ;   $P(R_{-1}) = P(R_{-1},-1) + P(R_{-1},0) + P(R_{-1},1)$
$P(0) = P(0,R_{-1}) + P(0,R_0) + P(0,R_1)$   ;   $P(R_0) = P(R_0,-1) + P(R_0,0) + P(R_0,1)$
$P(1) = P(1,R_{-1}) + P(1,R_0) + P(-1,R_1)$   ;   $P(R_1) = P(R_1,-1) + P(R_1,0) + P(R_1,1)$

Note that joint probabilities such as $P(1,R_0)$ with a "comma" between event terms is equivalent to the probability $P(1 \cdot R_0)$ of their intersection; it is written this way as a notational convenience for clarity. Also note that in the 1st equation for the total probability P(-1), the element "-1" remains the same in every term of the sum because this is a sum over y-values: y= $R_{-1}$, $R_0$, $R_1$ to find P(x = -1).

A formal proof of the Total Probability Formula for the binary signal is accomplished by noting that the universal set S=0∪0$^c$=0∪1 and hence we can write $R_1$=S·$R_1$=(0∪1)·$R_1$; taking the probabilities on both sides of this equation leads directly to $P(R_1) = P(R_1 \cdot 0) + P(R \cdot 1)$. Similarly, for a trinary we write S = -1∪0∪1 and $R_1$=S·$R_1$= (-1∪0∪1)·$R_1$ and again taking the probability yields the result for $P(R_1)$ given above. The extension to an n-ary signal with n distinct levels should be obvious as the universal set S can always be expressed as the union of ME and CE events.

## 4.1.7.3 Bayes' Rule

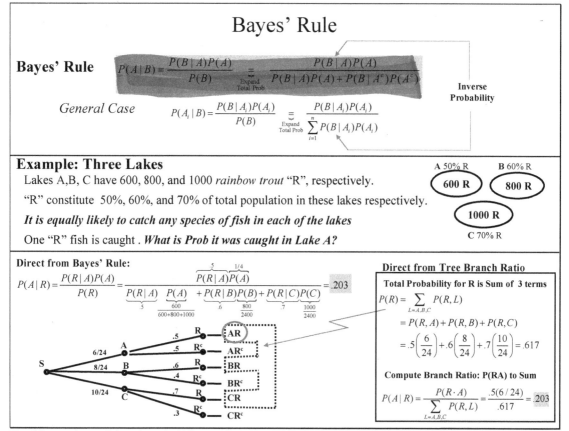

# Bayes' Rule

**Bayes' Rule**    $P(A\mid B)=\dfrac{P(B\mid A)P(A)}{P(B)}\underset{\substack{\text{Expand}\\\text{Total Prob}}}{\equiv}\dfrac{P(B\mid A)P(A)}{P(B\mid A)P(A)+P(B\mid A^c)P(A^c)}$    **Inverse Probability**

*General Case*    $P(A_i\mid B)=\dfrac{P(B\mid A_i)P(A_i)}{P(B)}\underset{\substack{\text{Expand}\\\text{Total Prob}}}{\equiv}\dfrac{P(B\mid A_i)P(A_i)}{\sum\limits_{i=1}^{n}P(B\mid A_i)P(A_i)}$

### Example: Three Lakes

Lakes A, B, C have 600, 800, and 1000 *rainbow trout* "R", respectively.

"R" constitute 50%, 60%, and 70% of total population in these lakes respectively.

*It is equally likely to catch any species of fish in each of the lakes*

One "R" fish is caught. **What is Prob it was caught in Lake A?**

A 50% R    600 R    B 60% R    800 R    1000 R    C 70% R

**Direct from Bayes' Rule:**

$$P(A\mid R)=\frac{P(R\mid A)P(A)}{P(R)}=\frac{\overbrace{P(R\mid A)}^{.5}\overbrace{P(A)}^{1/4}}{\underbrace{P(R\mid A)}_{.5}\underbrace{P(A)}_{\frac{600}{600+800+1000}}+\underbrace{P(R\mid B)P(B)}_{.6\;\frac{800}{2400}}+\underbrace{P(R\mid C)P(C)}_{.7\;\frac{1000}{2400}}}=.203$$

Tree:

S — 6/24 → A — .5 → R : AR
      A — .5 → R$^c$ : AR$^c$
    — 8/24 → B — .6 → R : BR
      B — .4 → R$^c$ : BR$^c$
    — 10/24 → C — .7 → R : CR
      C — .3 → R$^c$ : CR$^c$

**Direct from Tree Branch Ratio**

**Total Probability for R is Sum of 3 terms**

$$P(R)=\sum_{L=A,B,C}P(R,L)$$
$$=P(R,A)+P(R,B)+P(R,C)$$
$$=.5\left(\frac{6}{24}\right)+.6\left(\frac{8}{24}\right)+.7\left(\frac{10}{24}\right)=.617$$

**Compute Branch Ratio: P(RA) to Sum**

$$P(A\mid R)=\frac{P(R\cdot A)}{\sum\limits_{L=A,B,C}P(R,L)}=\frac{.5(6/24)}{.617}=.203$$

The ideas of conditional probability, total probability, and inverse probability are packaged together and are generally interpreted in terms of Bayes' Rule, which is obtained by writing joint probability P(AB) in two equivalent ways, *viz.*, P(A|B)P(B)=P(AB)=P(B|A)P(A). This expression is an identity independent of any interpretation given to it. Yet the interpretation given in the communication channel and the clinical testing examples is crucial to the way Bayes' Rule is applied.

The three lakes problem is a nice example of the inverse probability interpretation of Bayes' Rule. The problem states the actual number of rainbow trout "R" in each lake as A(600), B(800), C(1000), and the percentage of the population for "R" in each lake as 50% in A, 60% in B, and 70% in C. The problem asks "If a fisherman returns with an "R" fish, then what is the probability he went to lake A?" That is, we want the inverse probability P(A|R). The three *a priori* probabilities of an "R –fish" coming from each of lakes A, B, and C *prior* to having gone fishing (no data yet) are given by their relative proportion in each lake. There are a total of 600+800+1000=2400 "R –fish" in all three lakes, so we find P(A)= 600/2400, P(B)=800/2400, and P(C)=1000/2400. These numbers are placed on the three branches leading to tree nodes A, B, and C respectively. From each of these primary nodes, there are two branches leading to nodes R and R$^c$ ; we assign conditional values to each branch according to the stated percentages as follows: P(R|A)=.5, P(R|B)=.6, and P(R|C)= .7; the three complementary conditional values for the branches leading to R$^c$ are also shown in the tree. Either substituting these values directly into the Bayes' rule formula or multiplying the probabilities along the annotated tree paths, we arrive at the inverse probability P(A|R)=.203 which is down from it's *a priori* value P(A)=.25; this is because an R –fish favors the other two lakes with their larger *a priori* probabilities P(B) = .333 and P(C) = .417

## 4.2 Bayes' Probability Interpretation -a priori and a posteriori

# Bayes' Probability Interpretation -*a priori* and *a posteriori*

Binary Comm Signal - 2 Levels $\{0,1\}$  Binary Decision - $\{R_0, R_1\}=\{($ "0" rcvd , "1" rcvd$\}$

Channel Characteristics:

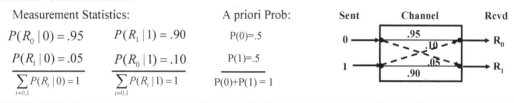

| Measurement Statistics: | | A priori Prob: | Sent | Channel | Rcvd |
|---|---|---|---|---|---|
| $P(R_0\|0)=.95$ | $P(R_1\|1)=.90$ | P(0)=.5 | | | |
| $P(R_1\|0)=.05$ | $P(R_0\|1)=.10$ | P(1)=.5 | | | |
| $\sum_{i=0,1}P(R_i\|0)=1$ | $\sum_{i=0,1}P(R_i\|1)=1$ | P(0)+P(1)=1 | | | |

**Absent measurements:** probability of a "0" or "1" = *a priori* values P(0)=P(1)=0.5
We have no reason to believe otherwise! A priori is judged likelihood of "0s" and "1s" in the data stream (in this case equally likely)

**Meas. #1** *changes our belief* by updating the *a priori* values according to Bayes' Rule

$$\overbrace{P(1\mid R_1)}^{a\ posteriori}=\frac{\overbrace{P(R_1\mid 1)}^{meas.\ statistic}\overbrace{P(1)}^{a\ priori}}{P(R_1\mid 1)P(1)+P(R_1\mid 0)P(0)}=\frac{(.90)\cdot(.5)}{(.90)\cdot(.5)+(.05)\cdot(.5)}=.947$$

**Note:** Measurement statistics are obtained by calibrating the detector on a source with *known* signals "0"&"1" and counting the # of correct detections in each case.

The interpretation of Bayes' rule in terms of our "belief" in the existence of a given state before and after measurement was discussed previously (Slide#3-6). For a binary communication signal Bayes' rule is interpreted as follows for the signal "1" (and analogously for the signal "0"):
(i) prior to measurements the signal is "believed" to be "1" with probability P(1)=1/2.
(ii) the conditional probability $P(R_1|1)$ is the "measurement statistic" that characterizes the accuracy with which a "1" is correctly detected; usually established by experiment using a known stream of digital "1"s
(iii) the inverse probability $P(1|R_1)$, is interpreted as the change in "belief" that the signal is indeed a "1" which results from a supporting measurement $R_1$; similarly for $P(1|R_0)$ and a non-supporting measurement $R_0$.
Thus we say that "a measurement updates our belief in a given state "1" from its *a priori* value P(1) =1/2 to a value obtained by multiplying P(1) by "belief multiplier". This multiplier $\{P(R_1|1) / P(R_1)\}$ is the ratio of conditional measurement statistic $P(R_1|1)$ for a "1" to $P(R_1)$, the total probability of the supporting measurement. The Bayesian update simply expresses this fact in the form $P(1|R_1)= \{P(R_1|1) / P(R_1)\}* P(1)$. Recall that the total probability of a supporting measurement $P(R_1)$ is written as $P(R_1) = P(R_1,0) + P(R_1,1)$, which may be re-expressed as $P(R_1) = P(R_1|0) P(0)+ P(R_1|1)P(1)$, which only involves known measurement statistics and a priori probabilities. The pair of update equations for "1" and "0" given $R_1$ then takes the form
$$P(1|R_1) = P(R_1|1) P(1) / [P(R_1|0) P(0)+ P(R_1|1)P(1)]$$
$$P(0|R_1) = P(R_1|0) P(0) / [P(R_1|0) P(0)+ P(R_1|1)P(1)]$$
Clearly, $P(1|R_1)$ will increase above it's *a priori* value of 1/2, while $P(0|R_1)$ decreases; their sum is unity.
The next few slides show how to make sequential Bayesian updates and also develop a method for easily computing these multiple updates using an additive logarithmic scale to replace the cumbersome multiplications and divisions.

## 4.2.1 Bayes' Rule and Multiple Measurement Updates

# Bayes' Rule and Multiple Measurement Updates

**Meas. #2** updates Meas. #1 using Definition of Conditional Probability

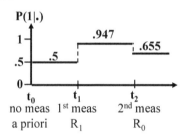

Meas.#1  Meas.#2

$$P(1 \mid R_1, R_0) = \frac{P(R_1, R_0, 1)}{P(R_1, R_0)} = \frac{\overbrace{P(R_0 \mid R_1, 1)}^{.10} \cdot \overbrace{P(R_1 \mid 1)}^{.90} \cdot \overbrace{P(1)}^{.5}}{\underbrace{P(R_0 \mid R_1, 0)}_{\substack{=P(R_0|0)=.95 \\ \text{indep meas } R_0, R_1}} \cdot \underbrace{P(R_1 \mid 0)}_{.05} \cdot \underbrace{P(0)}_{.5} + P(R_0 \mid R_1, 1) \cdot P(R_1 \mid 1) P(1)} = .655$$

**Total Probability Expansion** $P(R_1, R_0)$

$$P(R_1, R_0) = P(R_1, R_0, 0) + P(R_1, R_0, 1)$$
$$= P(R_0 \mid R_1, 1) \cdot P(R_1, 1) + P(R_0 \mid R_1, 0) \cdot P(R_1, 0)$$
$$= P(R_0 \mid R_1, 0) \cdot P(R_1 \mid 0) \cdot P(0) + P(R_0 \mid R_1, 1) \cdot P(R_1 \mid 1) P(1)$$

Note: measurements $R_1$ & $R_0$ are independent $\Rightarrow P(R_0 \mid R_1, 1) = P(R_0 \mid 1)$

**Sequential Updates**

**"Belief"** that "1" was sent prior to the 1st measurement is 0.5.
**Meas#1** $R_1$ confirms & thus increases our belief to .947
**Meas#2** $R_0$ disconfirms & thus decreases our belief to .655

The 1st measurement $R_1$ supports the state "1" leading to an increase in P(1) from 0.5 to P(1|R$_1$)=.947 as found on the last slide. Now consider a 2nd measurement R$_0$, which does not support the state "1". The new inverse probability of "1" conditioned on both measurements P(1|R$_1$,R$_0$) is obtained from the definition of conditional probability as P(R$_1$,R$_0$,1)/ P(R$_1$,R$_0$); using the concept of total probability the denominator is written as P(R$_1$,R$_0$) = P(R$_1$,R$_0$,0) + P(R$_1$,R$_0$,1), so the conditional probability is given by P(1|R$_1$,R$_0$) = P(R$_1$,R$_0$,1) / [P(R$_1$,R$_0$,0) + P(R$_1$,R$_0$,1)]. The joint probabilities in this expression are expanded by "chaining" the conditionals, as shown on the slide, and the denominator becomes P(R$_1$,R$_0$) = P(R$_0$|R$_1$,0)P(R$_1$|0)P(0) + P(R$_0$|R$_1$,1)P(R$_1$|1)P(1) = P(R$_0$|0)P(R$_1$|0)P(0) + P(R$_0$|1)P(R$_1$|1)P(1), The second equality follows because the measurements R$_0$ and R$_1$ are independent of one another which means P(R$_0$|R$_1$,0) = P(R$_0$|0) and P(R$_0$|R$_1$,1) = P(R$_0$|1). Substituting these known *a priori* and measurement statistics into the final expression on the slide yields P(1|R$_1$R$_0$) =.655.

The probability of the signal being a "1" is shown in the graph as a function of the measurement updates over time. Prior to any measurements the probability is .5 and remains at this value until the 1st "supporting" measurement R$_1$ occurs; the Bayesian update for R$_1$ increases the probability to .947 and it again remains unchanged until the 2nd "non-supporting" measurement R$_0$ occurs; the 2nd Bayesian update for R$_0$ decreases the probability to .655 where it remains waiting for further update measurements.

The next slide shows a tree for this update process and such a tree is the motivation behind the interpretation of the Bayesian update as an "inverse conditional probability."

## 4.2.1.1 Tree Computation of Inverse Probability $P(1|R_0 R_1)$

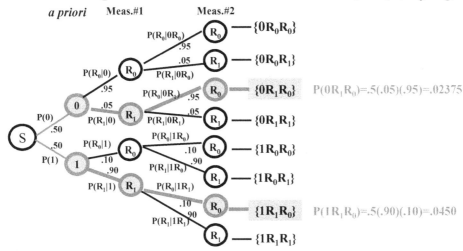

The two bolded red paths yield components of the *total probability equation*

$$P(R_1 R_0) = P(0R_1 R_0) + P(1R_1 R_0).$$

This identity expresses the total probability for the compound Event $R_1 R_0$ as a sum over the two tree path contributions through "0" and "1". Therefore, the "inverse probability" that the event $R_1 R_0$ emanated from "1" is simply the probability associated with that path over the total probability (sum of the two contributions). This yields a formula equivalent to the Bayesian update of the last slide, *viz.*,

$$P(1|R_1 R_0) = P(1R_1 R_0)/[P(0R_1 R_0) + P(1R_1 R_0)] = .0450/[.02375+.0450] = .655$$

The tree shows paths for all outcomes resulting from two measurements $R_0$ and $R_1$ on a digital signal have two possible values "0" and "1". If we concentrate on the two bolded red paths resulting from 1st measurement $R_1$ and 2nd measurement $R_0$ (same order as last slide), then we have two outcomes states denoted by $\{0R_1 R_0\}$ and $\{1R_1 R_0\}$ whose probabilities can be computed directly from the tree. Their sum yields the *total probability*

$$P(R_1 R_0) = P(0R_1 R_0) + P(1R_1 R_0),$$

Now it should be clear that, since there are *only two paths* for the measurements $R_1 R_0$ (*in that order*), the probability that the final state $\{1R_1 R_0\}$ emanated from node "1" is simply the ratio of $P(1R_1 R_0)$ to this sum; *i.e.,* the conditional probability is

$$P(1|R_1 R_0) = P(1R_1 R_0)/[P(0R_1 R_0) + P(1R_1 R_0)].$$

Similarly, the probability that the final state $\{0R_1 R_0\}$ emanated from node "0" is the ratio of $P(0R_1 R_0)$ to the same sum and yields the conditional probability

$$P(0|R_1 R_0) = P(0R_1 R_0)/[P(0R_1 R_0) + P(1R_1 R_0)].$$

Moreover, the sum $P(0|R_1 R_0) + P(1|R_1 R_0) = 1$, *i.e.*, the two conditional probabilities sum to unity (by construction) which is also easily verified by adding the two expressions above. More formally, we expand total probability as $P(R_1 R_0) = P(0|R_1 R_0) P(R_1 R_0) + P(1|R_1 R_0) P(R_1 R_0)$; canceling the common multiplier $P(R_1 R_0)$ on both sides of the equation yields the desired result: $1 = P(0|R_1 R_0) + P(1|R_1 R_0)$. Thus, the conditional probability $P(1|R_1 R_0)$ "*looks back into the tree*" to determine the likelihood that the pair of measurements $R_1 R_0$ were made on a signal with value "1" *rather than* "0". It is for this reason that it is called an "*inverse probability*".

Using the above formulas and computing values along the two tree paths $\{0R_1 R_0\}$ and $\{1R_1 R_0\}$ yields $P(1|R_0 R_1) = .0450/[.02375+.0450] = .655$ and $P(0|R_0 R_1) = .02375/[.02375+.0450] = .345$, which sum to unity as expected. [Note that the order of measurements is irrelevant, so for the reverse order $R_1 R_0$ we find the same results $P(1|R_1 R_0) = .655$ and $P(0|R_1 R_0) = .345$.]

### 4.2.1.2 Bayes' Rule Sequential Updates: "New" *a priori* Probability

## Bayes' Rule Sequential Updates: "New" *a priori* Probability

**Sequential Update Method:** Let the results of meas#1 represent "new" *a prioris,* and apply Bayes' for meas#2

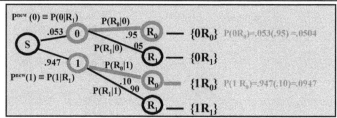

**"new"** $\quad P^{new}(1) = P(1 | R_1) = .947$

***a prioris*** $\quad P^{new}(0) = 1 - P(1 | R_1) = .053$

**Bayes' rule for meas#2 ("$R_0$")**

$$P(1 | R_0) = \frac{P(R_0 | 1) \cdot P^{new}(1)}{P(R_0 | 0) \cdot P^{new}(0) + P(R_0 | 1) \cdot P^{new}(1)} == \frac{(.10) \cdot (.947)}{(.95) \cdot (.053) + (.10) \cdot (.947)} = .655$$

**Total Probability Expansion Check of $P(1) = P(1R_0) + P(0R_0)$**

$$P(1) = P(1 | R_0) \cdot \underbrace{P(R_0)}_{=P(R_0,0)+P(R_0,1)} + P(1 | R_1) \cdot \underbrace{P(R_1)}_{=P(R_1,0)+P(R_1,1)}$$

$$= P(1 | R_0)\{P(R_0 | 0)P(0) + P(R_0 | 1)P(1)\} + P(1 | R_1)\{P(R_1 | 0)P(0) + P(R_1 | 1)P(1)\}$$

$$= .09524\{(.95)(.5) + (.1)(.5)\} + .947\{(.05)(.5) + (.90)(.5)\} = .5$$

Side Calculation $\quad P(1 | R_0) = \frac{P(R_0 | 1) \cdot P(1)}{P(R_0 | 0) \cdot P(0) + P(R_0 | 1) \cdot P(1)} = \frac{(.10) \cdot (.5)}{(.95) \cdot (.5) + (.10) \cdot (.5)} = .09524$

The **top panel** shows another method of computing multiple measurement updates is to actually do them sequentially by taking the result of the first measurement $R_1$ to be the "new *a priori*" for the second measurement $R_0$ and then performing the Bayesian update in the usual way. Accordingly, for this method we define $P^{new}(1) = P(1 | R_1) = .947$ and $P^{new}(0) = P(0 | R_1) = 1 - P(1 | R_1) = .053$, and then simply apply the Bayes' update formula for the single measurement $R_0$ using these new *a priori* values. The tree diagram in the top panel shows how the 1st measurement $R_1$ does not appear explicitly, but rather is accounted for by using the new *a priori* probabilities .053 and .947 emanating from the start. Using this tree, we can pick off the two terms in the Bayes computation to find the updated Bayesian probability for a "1" after the 2nd measurement $R_0$, *viz.*, $P(1 | R_0) = .655$, which agrees with our previous result.

The **bottom panel** computation is incidental to the sequential method discussed in the top panel. However, this total probability computation is of interest in a general context as it verifies explicitly that $P(1)$ *really is the sum* of the two joint probabilities $P(1\,R_0)$ and $P(1\,R_1)$ by explicitly computing the sum and showing it to be 0.5. This fact is most easily seen by visualizing the partitioning of the joint "sent-received" space as an overlay of the [0, 1]-sent plane with the [$R_0$, $R_1$]-received plane as shown in Slide#4-10. The overlay of these sent and received planes yields four distinct regions [1,$R_0$], [1,$R_1$], [0,$R_0$] and [0,$R_1$], and so by construction, it is clear that the "1-sent" partition is just the sum of [1,$R_0$] and [1,$R_1$] and the "0-sent" partition is the sum of [0,$R_0$] and [0,$R_1$]. The direct computation needed to actually show this result requires a little work as shown in the bottom panel of this slide.

## 4.2.2 Sequential Measurement Updates Using "Log-Odds Ratio"

# Sequential Measurement Updates Using "Log-Odds Ratio"

Logarithmic Scale "Adds" (similar to SNR expressed in dB)

$P(1|R_1)$ is the Bayesian update that a digital "1" was sent given a detection measurement $R_1$

**"Win/Lose" Odds Ratio**
Substitute Bayes' Rule

$$\frac{P(1|R_1)}{P(0|R_1)} = \frac{P(R_1|1)P(1)/P(R_1)}{P(R_1|0)P(0)/P(R_1)} = \frac{P(1)}{P(0)} \cdot \frac{P(R_1|1)}{P(R_1|0)}$$

**Note:** $0 = 1^c$

**Total Probability Identity**

$$P(R_1) = P(R_1 0) + P(R_1 1)$$

**Canceling $P(R_1)$**

$$P(R_1) = P(0|R_1) \cdot P(R_1) + P(1|R_1) \cdot P(R_1) \quad \textit{Yields} \quad 1 = P(0|R_1) + P(1|R_1)$$

Replacing $P(0|R_1)$ by $1 - P(1|R_1)$

$$\frac{P(1|R_1)}{P(0|R_1)} = \frac{P(1|R_1)}{1-P(1|R_1)} = \frac{P(1)}{P(0)} \cdot \frac{P(R_1|1)}{P(R_1|0)}$$

Taking the natural logarithm of the Odds Ratio yields

$$\ln\left[\frac{P(1|R_1)}{P(0|R_1)}\right] = \ln\left[\frac{P(1|R_1)}{1-P(1|R_1)}\right] = \ln\left[\frac{P(1)}{P(0)}\right] + \ln\left[\frac{P(R_1|1)}{P(R_1|0)}\right]$$

**Log-Odds Ratio "L"**

a posteriori log-odds ratio  a priori log-odds ratio  log-likelihood ratio incremental knowledge based on measurement

$$L_1 \quad = \quad L_0 \quad + \quad \Delta L$$

**Transformation**
$L_1$ to $P(1|R_1)$

$$L_1 \equiv \ln\left(\frac{P(1|R_1)}{1-P(1|R_1)}\right) \Rightarrow e^{L_1} = \frac{P(1|R_1)}{1-P(1|R_1)} \Rightarrow P(1|R_1) = \frac{e^{L_1}}{1+e^{L_1}}$$

The sequential Bayesian update method we have just discussed can be applied to any number of measurements. However, the procedure becomes a bit tedious for more than a few measurements; moreover, sometimes we wish to estimate the number of measurements required to obtain a certain target probability. Introduction of the log-odds ratio and its associated additive scale facilitates such calculations.

Consider the communication channel and start with the *a priori* probability $P(1)=1/2$ for a "1"; now use Bayes' Rule to update the *a priori* probability to the probability $P(1|R_1)$ conditioned on the measurement $R_1$. Dividing this result by its complementary probability $P(0|R_1)$ we obtain the "odds ratio" $P(1|R_1)/P(0|R_1)$ which expresses the ratio of a "win (1)" to a "loss (0)" given the measurement $R_1$. Expanding the numerator and denominator of this ratio using Bayes' Rule we find that the updated odds ratio can be expressed as the product of the *a priori* odds ratio $P(1)/P(0)$ and the measurement statistics ratio $P(R_1|1)/P(R_1|0)$ which is the likelihood of a *correct measurement* "$R_1$ given 1" to an *incorrect measurement* "$R_1$ given 0". Taking natural logarithms of both sides leads to the additive equation for the log-odds ratio shown in the boxed equation.

This equation expresses the updated log-odds ratio $L_1$ as the sum of the *a priori* log-odds ratio $L_0$ plus the increment $\Delta L$ determined by the log likelihood ratio of the measurement statistics. Thus the increment $\Delta L$ is the *strength of the measurement* characterized by how much more likely the *correct measurement* "$R_1$ given 1" is compared to the *incorrect measurement* "$R_1$ given 0". The logarithm of this ratio $\log[P(R_1|1)/P(R_1|0)]$ is positive for confirming ratios *greater than 1* and negative for *refuting* ratios *less than 1*; a ratio of *exactly 1* leads to a zero increment $\Delta L = \log[1] = 0$, which means the measurement has no effect at all. In the next slide we will revisit the communication sequential measurement update problem using the log-odds ratio and see how easy these updates become.

## 4.2.2.1 Log-Odds Ratio: Add/Subtract Measurement Information

# Log-Odds Ratio: Add/Subtract Measurement Information

Revisit Binary Comm Channel  $P(R_0 | 0) = .95$   $P(R_1 | 1) = .90$   $P(0)=.5$

Note:
$E = \text{"1"}$
$E^c = \text{"0"}$

$P(R_1 | 0) = .05$   $P(R_0 | 1) = .10$   $P(1)=.5$

Relation between $L_1$ and $P(1|R_1)$

$$L_1 \equiv \ln\left(\frac{P(1|R_1)}{1-P(1|R_1)}\right) \Rightarrow e^{L_1} = \frac{P(1|R_1)}{1-P(1|R_1)} \Rightarrow \qquad P(1|R_1) = \frac{e^{L_1}}{1+e^{L_1}}$$

$$\underbrace{\ln\left(\frac{P(1|R_1)}{P(0|R_1)}\right)}_{\equiv L_1} = \underbrace{\ln\left(\frac{P(1)}{P(0)}\right)}_{\equiv L_0} + \underbrace{\ln\left(\frac{P(R_1|1)}{P(R_1|0)}\right)}_{\equiv \Delta L_1}$$

Additive Meas Updates for L

Updates

$$L_{new} = L_{old} + \Delta L_{R_1} \qquad L_{old} = \ln\left(\frac{P(1)}{P(0)}\right) ; \ \Delta L_{R_1} = \ln\left(\frac{P(R_1|1)}{P(R_1|0)}\right)$$

| Meas#1: $R_1$ | Meas#2: $R_0$ | Alternate Meas#2: $R_1$ |
|---|---|---|
| $L_{old} = \ln\left(\frac{.5}{.5}\right) = 0$ <br><br> $\Delta L_{R_1} = \ln\left(\frac{.9}{.05}\right)$ <br><br> $= 2.8903$ <br><br> $L_{new} = 0 + 2.8903$ | $\Delta L_{R_0} = \ln\left(\frac{P(R_0|1)}{P(R_0|0)}\right) = \ln\left(\frac{.10}{.95}\right)$ <br><br> $= -2.25129$ <br><br> $L_{new} = L_{old} + \Delta L_{R_0}$ <br><br> $= 2.8903 + (-2.25129) = 0.63901$ | $\Delta L_{R_1} = \ln\left(\frac{P(R_1|1)}{P(R_1|0)}\right) = \ln\left(\frac{.90}{.05}\right)$ <br><br> $= +2.8903$ <br><br> $L_{new} = L_{old} + \Delta L_{R_1}$ <br><br> $= 2.8903 + 2.8903 = 5.7806$ |
| $P(1|R_1) = \frac{e^{2.8903}}{1+e^{2.8903}} = .947$ | $P(1|R_1 R_0) = \frac{e^{0.63901}}{1+e^{0.63901}} = .655$ | $P(1|R_1 R_1) = \frac{e^{5.7806}}{1+e^{5.7806}} = .997$ |

Revisiting the binary communication channel, we now compute updates using the log odds ratio which are additive updates. The update equation simply starts from the initial log odds ratio for the communication channel which is $L_{old}=\ln[P(1)/P(1^c)]= \ln(.5/.5)= 0$. There are two distinct measurement types $R_1$ and $R_0$; each adds an increment $\Delta L$ determined by its associated measurement statistics, *viz.*,

$R_1$: $\Delta L_{R1} =\ln[(P(R_1|1)/P(R_1|1^c)]=\ln(.90/.05) = +2.8903$  (positive "confirming")
$R_0$: $\Delta L_{R0} = \ln[(P(R_0|1)/P(R_0|1^c)]=\ln(.10/.95)= -2.25129.$ (negative "refuting")

The table illustrates how easy it is to accumulate the results of two measurements $R_1$ followed by $R_0$ by just adding the two $\Delta L$s to obtain

$L_{new}= 0+2.8903-2.25129=.63901,$

or alternately $R_1$ followed by $R_1$ to obtain

$L_{new}=0+2.8903+2.8903=5.7806.$

Note that the order of the measurements is obviously irrelevant since the order of the addition of increments $\Delta L_{R1}$ and $\Delta L_{R0}$ does not change the sum. Also note that after computing $P(1|R_1 R_0) =.655$, we can compute its complement based on the same measurements as $P(0|R_1 R_0)=1- P(1|R_1 R_0)=.345$ . These log odds ratios are converted to actual probabilities computing $P= e^{Lnew} / (1+ e^{Lnew})$ yielding .655 and .997 for the above two cases. If we want to find the number of $R_1$ measurements needed to give .99999 probability of "1" we need only convert .99999 to an $L =\ln[(.99999)/(1-.99999)] =11.51$ and divide the result by the increment per measurement $\Delta L= 2.8903$ to find 3.98 measurements , so 4 $R_1$ measurements are sufficient.

## 4.2.2.2 Mapping: Probability to Log-Odds Ratio

# Mapping: Probability to Log-Odds Ratio

**Sequential Updating of Probabilities**

**Transformation** of Probability Scale  P(L)

from  $L \in (-\infty, +\infty)$ to $P(L) \in [0,1]$

$$L = \ln\left[\frac{P}{1-P}\right]$$

$$P(L) = \frac{e^L}{1+e^L}$$

**Computation Steps:**

| | |
|---|---|
| **1)** Start with old value of L | $L_{old} = \ln\left(\frac{P(1)}{P(0)}\right)$ |
| **2)** Compute Incremental Knowledge $\Delta L$ from measurement  $R_1$  in terms of the Log-likelihood ratio) | $\Delta L_{R_1} = \ln\left(\frac{P(R_1 \mid 1)}{P(R_1 \mid 0)}\right)$ |
| **3)** Update the L-value | $L_{new} = L_{old} + \Delta L_{R_1}$ |
| **4)** Convert to Probability using inverse transformation P(L) | $P(1 \mid R_1) = \frac{e^{L_{new}}}{1+e^{L_{new}}}$ |

The log-odds ratio expression $L(P) = \ln[P/(1-P)]$ defines L as a function of P and serves to convert probabilities into their associated log-odds ratios in preparation for the simple additive form of Bayesian updates applicable to any number of sequential measurements.  Once done, we need to convert the final log-odds ratio value "L" back into an actual probability value "P."    This is accomplished by the inverse transformation $P(L) = e^L / (1+e^L)$ which maps values of L in the interval $(-\infty, +\infty)$ into values of P in the restricted interval [0,1] as shown in the plot of the upper panel.

A summary table of the computation steps needed to perform sequential updates, using the log-odds ratio, is given in the lower panel.  The expressions in this table are for updates on the binary digit "1"; there are analogous steps for the digit "0.  Steps shown are first a conversion of the *a priori* ratio to a log-odds ratio $L_{old}$, then conversion of the measurement statistics ratio to an incremental knowledge $\Delta L$, followed by their addition to form $L_{new}$ (repeated as many times as there are measurements), and finally an inverse transformation P(L) to convert the result back into a probability.

## 4.2.3  Bayes' Rule: Inverse Conditional (Antecedent) Probabilities

# Bayes' Rule: Inverse Conditional (Antecedent) Probabilities

**Problem Statement:**

- Urn contains r - Red and b-Blue balls  N= r + b
- **n** balls are drawn sequentially     $n \leq r + b = N$
- Given: **$k\,blue$** *drawn  -  Event* $B_k$
- Find Cond Prob that *$1^{st}$ ball drawn is blue*   $P(B_1|B_k)$

Urn: N = r + b       Urn*: n = (n-k)+k

**Argue1:** *Put in new urn  U\* which has k  blue & (n-k) red balls =>*  $P(B_1|U^*) = k/n$

**1)** *a priori* $1^{st}$ ball is blue   $P(B_1) = \dfrac{b}{N} = \dfrac{b}{r+b}$

**2)** *a priori* k blue   $P(B_k) = \dfrac{\binom{b}{k}\binom{r}{n-k}}{\binom{r+b}{n}}$

$N\ choose\ n$

**3)** k blue given $1^{st}$ is blue   $P(B_k|B_1) = \dfrac{\binom{b-1}{k-1}\binom{r}{n-k}}{\binom{r+b-1}{n-1}}$

$N-1\ choose\ n-1$

Cond.    Bayes'
Prob.    Rule

$$P(B_1|B_k) = \frac{P(B_1 B_k)}{P(B_k)} = \frac{P(B_k|B_1)P(B_1)}{P(B_k)}$$

$$= \frac{\dfrac{\binom{b-1}{k-1}\binom{r}{n-k}}{\binom{r+b-1}{n-1}} \cdot \left(\dfrac{b}{r+b}\right)}{\dfrac{\binom{b}{k}\binom{r}{n-k}}{\binom{r+b}{n}}} = \frac{\binom{r+b}{n}}{\binom{b}{k}\binom{r}{n-k}} \cdot \frac{\binom{b-1}{k-1}\binom{r}{n-k}}{\binom{r+b-1}{n-1}} \cdot \left(\frac{b}{r+b}\right) = \frac{\binom{b-1}{k-1}}{\binom{b}{k}} \cdot \frac{\binom{r+b}{n}}{\binom{r+b-1}{n-1}} \cdot \left(\frac{b}{r+b}\right)$$

$$= \frac{\dfrac{(b-1)!}{(k-1)!(b-k)!}}{\dfrac{b!}{k!(b-k)!}} \cdot \frac{\dfrac{(r+b)!}{n!(r+b-n)!}}{\dfrac{(r+b-1)!}{(n-1)!(r+b-n)!}} \cdot \left(\frac{b}{r+b}\right) = \frac{k}{b} \cdot \frac{r+b}{n} \cdot \left(\frac{b}{r+b}\right) = \boxed{\frac{k}{n}}$$

**Argue2)** *There are exactly  k of n equally likely tree graphs that have a blue drawn $1^{st}$. Thus* $P(B_1|B_k) = k/n$

---

 n balls are drawn from an urn containing b-blue and r-red (r+b=N) resulting in " k-blue and (n-k)-red. "  We are asked to predict backwards to find the probability that the first ball drawn was blue labeled event $B_1$; that is, find the inverse Bayes conditional $P(B_1|B_k)$ that the 1st ball drawn is blue given that k of the n drawn are blue. Bayes' rule (boxed equation) gives $P(B_1|B_k)$ in terms of the "measurement statistic" $P(B_k|B_1)$, the *a priori* $P(B_1)$ and the total probability $P(B_k)$.  These quantities are computed in the line above Bayes' rule and substitution yields, after some algebraic manipulation, the desired result $P(B_1|B_k) = k/n$ at the bottom of the slide.

The calculations of these three quantities can be argued as follows: The *a priori* probability $P(B_1)$ that B is drawn first is simply the ratio of blue to the total or b/(r+b).  The probability $P(B_k|B_1)$ of drawing  k-blue given the $1^{st}$ draw is blue, is obtained by first drawing k-1 more blues from the remaining b-1 blue or $^{b-1}C_{k-1}$ ways; next draw n-k red from  r reds or $^{r}C_{n-k}$ ways; finally dividing their product by the total #ways to draw n-1 from r+b-1, we find the conditional $P(B_k|B_1)= {}^{b-1}C_{k-1}\,{}^{r}C_{n-k} / {}^{b+r-1}C_{n-1}$ .  The probability of drawing k blue and n-k reds from b blues and r reds is given by $P(B_k) = {}^{b}C_k\,{}^{r}C_{n-k} / {}^{b+r}C_n$  (see Slide#2-22 identity (6) hypergeometric expansion terms).

There are two combinatorial arguments that yield the same result:

**i)** If upon drawing each of the balls they are placed in a second urn U*,  then this urn will have exactly k blue balls and (n-k) red balls in it.  Before we take the $1^{st}$ draw from U* the "new *a priori* probability" is simply the ratio (#blue)/(#blue + #red) in that urn.  Thus, since the urn U* has k blue and (n-k) red (by construction) the ratio is given by $P_{U^*}(b)= k/(k+(n-k))= k/n$.  Thus equating the new *a priori* (See Slide# 4-16) to the conditional probability of the original problem we have $P(B_1|U^*)= P_{U^*}(b)= k/n$ . Note that the conditioning imposed by U* is exactly that of the original problem, *i.e.*, n draws result in k blue and n-k red.

**ii)** The n draws from the original urn can be represented by a tree.  Clearly any ball (blue or red) could be drawn first and hence there are n different trees each of which has a different ball drawn first.  In k of these n trees a blue ball is drawn first and hence the probability of a blue ball being drawn first is simply k/n .

# Probability Types, Interpretations, and Visualizations

## 4.2.3.1 Bayes' Rule Examples

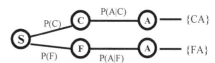

# Bayes' Rule Examples

- Ex. 1) French F or Chemistry C
- $P(A|F)=1/2$ ; $P(A|C) =2/3$
- Flips Fair Coin to Decide which to take
- $P(C)=1/2$;  $P(F)=1/2$
  - Find Prob receives an A in Chemistry

$$P(AC) = P(A|C)P(C) = \frac{2}{3}\frac{1}{2} = \frac{1}{3}$$

**Partial Tree**

{CA}

{FA}

If receives an A what is Prob took Chemistry

$$P(C|A) = \frac{\overbrace{P(A|C)}^{2/3}\overbrace{P(C)}^{1/2}}{\underbrace{P(A|C)P(C)}_{2/3\quad 1/2}+\underbrace{P(A|F)P(F)}_{1/2\quad 1/2}} = \frac{4}{7}$$

---

- Ex. 2) Urn   8 Red,  4 White  Balls
- Draw two without replacement
- Find prob of two reds  P(RR)

$P(R_2|R)$

Drw#1   Drw#2

**Partial Tree**

{R1 R2}

{R1 W1}

$$P(R_1R_2) = \frac{\binom{8}{2}\binom{4}{0}}{\binom{12}{2}} = \frac{(8\cdot 7)/2}{(12\cdot 11)/2} = \frac{14}{33}$$

$$P(R_1R_2) = \underbrace{P(R_2|R_1)}_{7/11}\underbrace{P(R_1)}_{8/12} = \frac{14}{33}$$

✱ WHAT IS THE POINT OF THE TREE ?

Here are two more simple examples using conditional probability.

In the first, a **student flips a coin** to determine which course to take French F or Chemistry C. The conditional probabilities to receive an A in each course are given as P(A|F) and P(A|C). Following the tree from start to C and then to Grade A yields AC and multiplying terms along the path represents the probability P(AC)= P(A|C) P(C); similarly for the path to AF. Given the student receives an A, the probability he took Chemistry C is expressed as the inverse conditional probability P(C|A) and solved using Bayes' rule.

In the second example, an **urn contains {8R , 4W}** and two are drawn without replacement. We are asked to find probability of drawing 2R. Both a direct calculation and the associated tree diagram illustrate the computation of joint probability in terms of the conditional and a priori probabilities according to  $P(R_1R_2)=P(R_2|R_1)P(R_1)$. A third technique effectively places the R and W balls into separate urns and then computes the #ways to choose 2 of 8 from the red urn $^8C_2$ and the #ways to choose 0 of 4 from the white urn $^4C_0$. The product of these two divided the total #ways to choose 2 of 12 in the original urn $^{12}C_2$ gives the desired probability P(2R) = $^8C_2\,^4C_0 /\,^{12}C_2$.

The binomial identity (6) on Slide#2-22, can shed some light on this problem by considering the expansion  $^{12}C_2=^{8+4}C_2=^8C_0\,^4C_2+^8C_1\,^4C_1+^8C_2\,^4C_0$.  The expansion effectively performs the "urn separation" by taking "2 from W urn" or "1 from each urn" or "2 from R urn". These three terms enumerate all outcomes and upon division by $^{12}C_2$ yields 1= $^8C_0\,^4C_2 /^{12}C_2+ ^8C_1\,^4C_1 /^{12}C_2 + ^8C_2\,^4C_0/^{12}C_2$. The third term is the one we have calculated and the three take together constitute a hypergeometric probability distribution.

## 4.2.3.2 Bayes' Update - Monte Hall Problem

# Bayes Update - Monte Hall Problem

**Guest at a Quiz Show is presented with 3 Doors and chooses one but does not open it.**

**The host knows where the grand prize Car (C) is and opens a door to reveal a Goat (G).**

**The guest is then asked if he wants to keep the door he picked or switch to the remaining door.**

**What is his best option?**

**1) Marylin vos Savant Solution:**

**Table shows all possible permutations behind the three doors**

**C = Car, G= Goat ; Host Choices Boxed**

| Guest | Host | | | | |
|---|---|---|---|---|---|
| $D_1$ | $D_2$ | $D_3$ | KEEP | SWITCH |
| C | G | G | C | G |
| G | C | G | G | C |
| G | G | C | G | C |
| | Prob (C) = | | 1/3 | 2/3 | |

**2) Tree/ Bayesian Update**

i) **Guest chooses Door#1: $G_1$**
ii) **Host chooses Door#3: $H_3$**
iii) **Guest Either** *Keep* $D_1$
or *Switch* to $D_2$

**Keep $D_1$** $P(C_1|H_3) = P(H_3|C_1) P(C_1) / P(H_3) = (1/2*1/3)/(1/3+1/6) = 1/3$

**Switch $D_2$** $P(C_2|H_3) = P(H_3|C_2) P(C_2) / P(H_3) = (1*1/3)/(1/3+1/6) = 2/3$

[Note: $P(C_3|H_3) = 0$ *as Host never chooses Door with Car*]

$P(H_3) = P(H_3|C_1) P(C_1) + P(H_3|C_2) P(C_2)$

$= 1/2 * 1/3 + 1 * 1/3 = 3/6$

**Event $G_1$ = Guest chooses Door #1**
**Event $C_1$ = Car behind Door #1 , etc.**
$P(C_1) = P(C_2) = P(C_3) = 1/3$ *equally probable*
**Event $H_3$ = Host chooses Door#3**

*Generalize n-doors*
$P(Keep) = 1/n$
$P((n-2)-Sw) = 1-1/n$
$P(1-Sw) = (n-1) / [n(n-2)]$

The Monte Hall Problem derives from a popular quiz show back in the 1970s in which the host presents three doors to a guest indicating that there is a brand new car C behind one and a goat G behind the other two. The guest is asked to choose one door without opening it and the host then opens one of the two remaining doors to reveal a goat! The guest is then asked to decide whether to keep the door originally chosen or switch to the other unopened door. What is the guest's best decision, *i.e.*, the one with the highest probability of winning the car? A New York Times Parade magazine column by "Marilyn vos Savant" gave a very simple solution which showed that switching to the other door was the best strategy. This led to a very heated controversy involving well known mathematicians, physicists, and others casting doubt on her solution indicating that it should not make any difference whether the guest keeps or switches doors. In the end it turned out that Marilyn, who was an extremely bright woman, was in fact correct. .

The *upper panel* gives Marilyn's solution table displaying the three possible arrangements of the two goats (G) and one car (C) behind the three doors as rows under the column headers $D_1$, $D_2$, $D_3$. This table assumes that the Guest chooses $D_1$, and the Host chooses between doors $D_2$ and $D_3$ to reveal a goat G as shown (surrounded by a box) in each row. The last two columns label the outcomes for the Guest's choice either to KEEP $D_1$ or to SWITCH to the remaining unopened door. Row#1 illustrates that the Host can choose either $D_2$ or $D_3$ since each has a G behind it and the outcome is that the Guest wins the car C if he decides to KEEP, but gets the goat G if he decides to SWITCH. The KEEP and SWITCH outcomes for the remaining two rows are tabulated and the resulting probability of winning the car C is 1/3 for KEEP and 2/3 for SWITCH. The *bottom panel* shows a simple tree representation that also assumes the Guest chooses door $D_1$. Independent of the Guest's choice, the car is equally likely to be behind each of the three doors, so $P(C_1) = P(C_1) = P(C_1) = 1/3$. Now if the car is behind $D_1$, (event $C_1$), then the Host can choose either $D_2$ or $D_3$, leading to events $H_2$ or $H_3$ illustrated on the tree branches with conditional probability $P(H_3|C_1) = P(H_2|C_1) = 1/2$. On the other hand, if the car is behind either door $D_2$.or $D_3$, the Host has only one choice in each case, so $P(H_3|C_2) = 1$ or $P(H_2|C_3) = 1$ respectively. Thus assuming the host chooses $H_3$ we can use Bayes' Rule to look back into the tree and find the probability $P(C_1|H_3) = P(H_3|C_1)P(C_1)/P(H_3)=1/3$ which is the win probability if the Guest *keeps* his original door $D_1$ choice and $P(C_2|H_3) = P(H_3|C_2)P(C_2)/P(H_3)=2/3$ which is the win probability if the Guest *switches* to $D_2$. This agrees with the simple tabular method of the top panel. For n-doors $P(Keep)+P((n-2)-Sw) =1$ and since $P(keep)=1/n$, we have $P(1-Sw)=( 1-1/n)/(n-2) = (n-1)/n/(n-2)$.

## 4.2.4 Conditional "Chains": Aces into 4 Piles

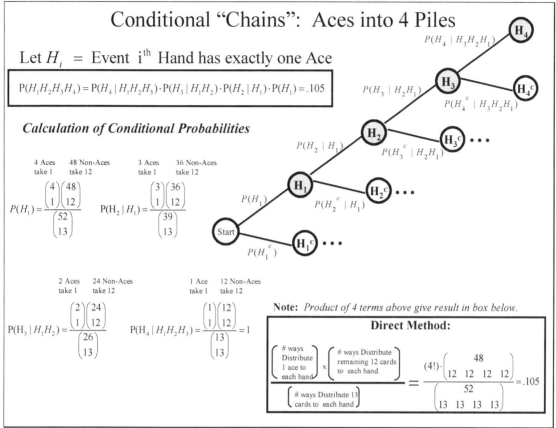

Conditional "Chains": Aces into 4 Piles

Let $H_i$ = Event i$^{th}$ Hand has exactly one Ace

$$P(H_1 H_2 H_3 H_4) = P(H_4 \mid H_1 H_2 H_3) \cdot P(H_3 \mid H_1 H_2) \cdot P(H_2 \mid H_1) \cdot P(H_1) = .105$$

*Calculation of Conditional Probabilities*

4 Aces   48 Non-Aces      3 Aces   36 Non-Aces
take 1     take 12        take 1     take 12

$$P(H_1) = \dfrac{\binom{4}{1}\binom{48}{12}}{\binom{52}{13}} \qquad P(H_2 \mid H_1) = \dfrac{\binom{3}{1}\binom{36}{12}}{\binom{39}{13}}$$

2 Aces   24 Non-Aces      1 Ace   12 Non-Aces
take 1     take 12        take 1    take 12

$$P(H_3 \mid H_1 H_2) = \dfrac{\binom{2}{1}\binom{24}{12}}{\binom{26}{13}} \qquad P(H_4 \mid H_1 H_2 H_3) = \dfrac{\binom{1}{1}\binom{12}{12}}{\binom{13}{13}} = 1$$

**Note:** *Product of 4 terms above give result in box below.*

**Direct Method:**

$$\frac{\left(\begin{array}{c}\text{\# ways}\\\text{Distribute}\\\text{1 ace to}\\\text{each hand}\end{array}\right) \times \left(\begin{array}{c}\text{\# ways Distribute}\\\text{remaining 12 cards}\\\text{to each hand}\end{array}\right)}{\left(\begin{array}{c}\text{\# ways Distribute 13}\\\text{cards to each hand}\end{array}\right)} = \frac{(4!) \cdot \binom{48}{12\;12\;12\;12}}{\binom{52}{13\;13\;13\;13}} = .105$$

Four hands are dealt from a deck of 52 cards and we are asked to find the probability that each hand has "exactly" one ace? A tree is very helpful in analyzing this problem. Let's define $H_1$ to be the event that hand#1 has exactly one ace; $H_2$ the event that hand#2 has exactly one ace, and similarly for $H_3$, and $H_4$. In terms of these events, we seek the joint probability $P(H_1 H_2 H_3 H_4)$ and this easily decomposes as a product of conditional probabilities as one travels along the upper tree path to the joint state $\{H_1 H_2 H_3 H_4\}$.

From the tree start position we have two branches leading to the ME and CE states $H_1$ and $H_1^c$ and the paths are labeled by the *a priori* probabilities $P(H_1)$ and $P(H_1^c)$; then to $H_2$ and $H_2^c$ with paths labeled by the conditional probabilities $P(H_2|H_1)$ and $P(H_2^c|H_1)$; then to $H_3$ and $H_3^c$ with path conditional probabilities $P(H_3|H_2 H_1)$ and $P(H_3^c|H_2 H_1)$; finally to $H_4$ and $H_4^c$ with path conditional probabilities $P(H_4|H_3 H_2 H_1)$ and $P(H_4^c|H_3 H_2 H_1)$. The probability of exactly 1 ace in hand#1, $P(H_1)$ is found by selecting 1 of 4 aces and 12 of 48 non-aces and then dividing by $^{52}C_{13}$ to give $P(H_1) = {}^4C_1 * {}^{48}C_{12} / {}^{52}C_{13}$. Note that the event $H_1$ says nothing about the distribution of the remaining 3 aces in the other hands; for example, hand #3 could have all 3 of the remaining aces. Given $H_1$ the conditional probability for $H_2$ is obtained by taking 1 of 3 remaining aces and 12 of 36 remaining non-aces and dividing by $^{52}C_{13}$ to give $P(H_2|H_1) = {}^3C_1 * {}^{36}C_{12} / {}^{52}C_{13}$ ; similarly for $H_3$: $P(H_3|H_2 H_1) = {}^2C_1 * {}^{24}C_{12} / {}^{52}C_{13}$ and for $H_4$: $P(H_4|H_3 H_2 H_1) = {}^1C_1 * {}^{12}C_{12} / {}^{52}C_{13}$ . We note that these conditional probabilities are just terms of different Hypergeometric expansions, *e.g.*, $^{52}C_{13} = {}^{4+48}C_{13} = {}^4C_0\,{}^{48}C_{13} + {}^4C_1\,{}^{48}C_{12} + {}^4C_2\,{}^{48}C_{11} + {}^4C_3\,{}^{48}C_{10} + {}^4C_4\,{}^{48}C_9$; $^{39}C_{13} = {}^{3+36}C_{13} = {}^3C_0\,{}^{36}C_{13} + {}^3C_1\,{}^{36}C_{12} + {}^3C_2\,{}^{36}C_{10} + {}^3C_3\,{}^{36}C_{10}$; $^{26}C_{13} = {}^{2+24}C_{13} = {}^2C_0\,{}^{24}C_{13} + {}^2C_1\,{}^{24}C_{12} + {}^2C_2\,{}^{24}C_{10}$.

The inset box shows an equivalent calculation that distributes 12 of the 48 non-aces to each hand with a multinomial $^{48}C_{12\,12\,12\,12}$ and then multiplies it by 4! (the number of ways 4 aces can be distributed one to each hand). Dividing by the total number of ways to distribute the cards $^{52}C_{13\,13\,13\,13}$ yields the same result .105 .

## 4.2.4.1 Aces into 4 Piles – Alternate Event Space

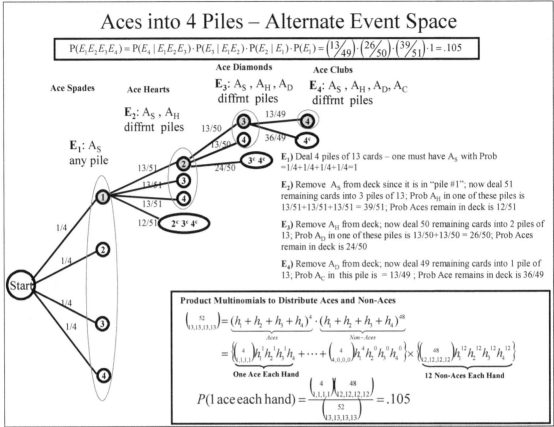

This method is discussed by Ross [Ref. 1], but remains very cryptic without an illustration. The events E define increasingly restrictive distributions of the 4 aces. They are not the same as the previous events H stating, *e.g.*, $H_1$ "hand#1 has exactly one ace".

**The event $E_1$:** states the ace of spades "$A_S$ is in any pile" and hence state $E_1$ is the group of primary nodes which is the "whole space."

**The event $E_2$:** states that "$A_S$ and $A_H$ are in different piles." The partial tree illustrates the case for which the $A_S$ is in hand#1; so following the branch emanating from "1", we see that $E_2$ states that the $A_H$ is either in the union $2 \cup 3 \cup 4$ or in its complement $(2 \cup 3 \cup 4)^c = 2^c\ 3^c\ 4^c$ (not any). Event $E_2$ is the group of 3 shown in the "Ace Hearts column" of the tree.

**The event $E_3$:** states that "$A_S$, $A_H$ and $A_D$ are in different piles" so if we now assume that the $A_H$ is in hand#2 then $E_3$ means that the $A_D$ is either in the union $3 \cup 4$ or in its complement $(3 \cup 4)^c = 3^c\ 4^c$ (not either). Event $E_3$ is the group of 2 shown in the "Ace Diamonds column" of the tree.

**The event $E_4$:** states that "$A_S$, $A_H$, $A_D$, and $A_C$ are in different piles" so if we now assume that the $A_D$ is in hand#3, then $E_4$ means that the $A_C$ is either in 4 or in its complement $4^c$.

An approximate solution is found by noting that the conditional probabilities for the increasingly restrictive states are reduced by approximately 1/4 at each node of the tree, *viz.*, $P(E_1)=1$, $P(E_2|E_1)\sim 3/4$, $P(E_3|E_1E_2)\sim 1/2$, $P(E_4|E_1E_2E_3)\sim 1/4$; their product is $1*3/4*1/2*1/4 = .09375$. Computing the actual conditionals along the path and taking their product yields the joint probability $P(E_1E_2E_3E_4)=.105$. Also note that we followed the "partial" tree for a specific order of hands #1,#2,#3,#4, but we could just as well have chosen a different order and obtained the same result. [Note that if we disregard everything but aces and assume they are independently assigned to four hands (which is not strictly true since they are dealt with other cards in the hand), then we can write down a multinomial $(h_1+h_2+h_3+h_4)^4$ which yields $4^4$ permutations and 35 groups. The boxed equation at bottom of slide shows, a product of two multinomials distributing respectively the 4 aces and the 52 non-Aces among the 4 hands. For 1 ace to each hand we choose the term $^4C_{1,1,1,1}$ and divide by the number of ways to distribute just the aces *i.e.*, $4^4$, to find P(1 ace each hand) = $^4C_{1,1,1,1}/4^4 = 4!/4^4 =.09375$. Taking into account the product multinomial yields the exact answer $^4C_{1,1,1,1}\ ^{48}C_{12,12,12,12} / \ ^{52}C_{13,13,13,13} =.105$]

# 5 Independent Sets of Events and Repeated Trials

# Independent Sets of Events & Repeated Trials

## 5.1  Independence of Events

# Independence of Events

**Two events A & B** are independent if they do not influence one another

**Definition:** $P(AB) = P(A) \cdot P(B)$   Joint probability factors into product of probabilities

(1a)  $P(A \mid B) = P(A)$

Pr("A given B")= Pr(A)

*i.e.,* B is irrelevant

Symmetric A⇔B

Consequence of Defn:

$P(A \mid B) = \dfrac{P(AB)}{P(B)} = \dfrac{P(A)P(B)}{P(B)} = P(A)$

(1b) $P(B \mid A) = P(B)$

(2a) $P(B \mid A) = P(B \mid A^c) = P(B)$

Symmetric A ⇔ B

occurrence or non-occurrence

of "other" is irrelevant

(2b) $P(A \mid B) = P(A \mid B^c) = P(A)$

**Ex. #1:** *Pull "Ace of Spades"* (AS) from Card Deck   $\underset{=1/52}{P(AS)} \overset{?}{=} \underset{=4/52}{P(A)} \cdot \underset{=13/52}{P(S)} = \frac{1}{52}$   **Yes Indep**

**Ex. #2:** *Flip Coin Twice*

4 Outcomes $\{H_1H_2, H_1T_2, T_1H_2, T_1T_2\}$

$\underset{=1/4}{P(H_1T_2)} \overset{?}{=} \underset{=2/4}{P(H_1)} \cdot \underset{=2/4}{P(T_2)} = \frac{1}{4}$   **Yes Indep**

**Ex. #3:** *Many Independent Coin Flips*

*All Heads:* $H_1H_2H_3H_4H_5 \cdots H_n$

$P(T \mid H_1H_2H_3H_4 H_5 \cdots H_n) = P(T) = 0.5$

$P(H \mid H_1H_2H_3H_4 H_5 \cdots H_n) = P(H) = 0.5$

**Not "Due for a Tail"**

**Each Flip Indep**

A pair of events are defined to be independent *if and only if* their joint probability P(A,B) factors into the product P(A)∗P(B).  This definition leads to a number of equivalent expressions of independence stating that the conditional probability of event A given either B or its complement $B^c$ reduces to the *a priori* value P(A), *viz.,*  P(A|B) = P(A|B^c) = P(A).  The primary definition for independence is symmetric and so too are the conditional versions with A and B swapped.  All expressions of independence state that the occurrence of one event (A) is irrelevant to the occurrence or non-occurrence of the other event (B); such a *statement of indifference* is what we intuitively understand as independence.  Thus it will always be the case that, if the two parameters are free to range over their sets of values without constraints, then events described by these parameter pairs are necessarily independent; hence, for independent events their joint probability equals the product of their individual probabilities.  We discuss two specific examples, a deck of cards and a fair coin flipped twice.

**For a deck of 52 cards** labeled by two distinct sets of parameters (i) numbers: 1,2, ..., 13, and (ii) suits {H,S,C,D}, the probability for the joint parameter event A·S ("ace of spades") factors into the product of their probabilities, *viz.,*  P(A·S) = P(A)∗P(S).  For the *ace of spades* we reason that since it is a unique card, its joint state A·S has probability 1/52, whereas the probability of an ace is 1/13 and that of a spade is 1/4, so substitution explicitly verifies independence 1/52 = (4/52)*(13/52).  A similar argument can be given for each and every card, thus verifying independence of the parameters *suit* and *number*.  This also agrees with our intuitive understanding that being a *heart* has no influence upon what *number* is assigned, and *vice-versa*, choosing a *number* does not bias our choice of *suit* in any manner.

**Flipping a Fair Coin Twice** is another example that blends in well with our intuitive understanding of independence.  Since there are 4 possible *joint outcomes*, the specific output $H_1T_2$ has probability 1/4, while the probability of a head or tail on any flip is 1/2; thus we verify independence by substituting these values into P($H_1T_2$) = P($H_1$)∗P($T_2$) to yield 1/4 = (1/2)∗(1/2).

**After flipping a Long sequence of Heads HHHHH** ⋯ **H** we are not **"Due for a Tail."** as some gamblers would like to think.  The probability for a head or a tail on the next flip is always the same for a fair coin. The clear meaning of independence is that previous trials have absolutely no influence on the very next flip!! They are all independent events.

# Independent Sets of Events & Repeated Trials

## 5.1.1 Independent Events for 4-Sided Dice

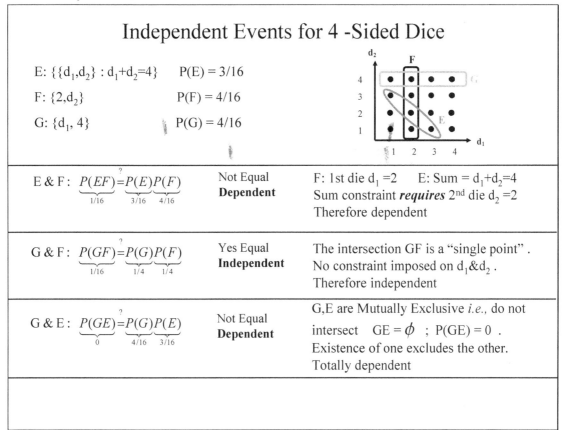

A more interesting example of event independence is illustrated in the sample space generated by a pair of 4-sided dice illustrated in this slide. We have overlaid several events on the standard $d_1$-$d_2$ coordinate representation as follows: event E ($d_1+d_2=4$) is the elliptical shape (red) running diagonally, event F ($d_1=2$) is the vertical rectangular shape (black), and G($d_2=4$) is the horizontal rectangular shape (green). The probabilities for each of these sets and their intersections are easily found by counting the equally likely "dots-elements" in each set and dividing by 16.

The factoring definition of independence is applied to the three joint events EF, GF, and FE and the tests show that G *and* F are *independent*, while both E *and* F and G *and* E are *dependent*. The fact the G *and* F are independent should be obvious from the fact that they range over two different (orthogonal) parameters $d_1$ for G and $d_2$ for F. They even "look orthogonal" and indeed they are verified to be independent by calculations on the slide. The dependence between E and F results from the constraints imposed on parameters $d_1$ and $d_2$ by the definitions of events E ($d_1+d_2=4$) and F($d_1=2$). Thus, since $d_1=2$ and their sum is 4, the only value for $d_2$ is 2 which constrains their intersection to the single point (2,2) as shown in the figure. The *dependence* between the non-intersecting sets G *and* E seems *counter-intuitive* at first, but can be made clear if we recognize that being in G "influences strongly" whether it is in E as well; *i.e.*, because of their null intersection an element in G cannot also be in E.

## 5.1.2 Disjoint *vs.* Independent, Alternate Forms, Conditional Independence

# Disjoint vs Independent, Alternate Forms, Conditional Independence

- **ME and Independent** are not the same

  Last Slide: Events G & E are ME $\qquad G \cdot E = \phi \implies P(GE) = 0 \neq \underbrace{P(G)}_{\neq 0} \cdot \underbrace{P(E)}_{\neq 0}$

  but they are **not independent**

  G & E are "totally dependent"; *i.e.,* one excludes the other

  *IF DISJOINT (ME) : P(AB)=0*

  **Alternate Forms**

- **A & B independent** $\qquad P(AB) = P(A) \cdot P(B)$ $\qquad$ 1) $P(AB^c) = P(A) \cdot P(B^c)$

  **Proofs:** $\qquad\qquad\qquad\qquad\qquad\qquad\qquad\qquad$ 2) $P(A^c B) = P(A^c) \cdot P(B)$

  1) $P(A) = P(A(B \cup B^c)) = P(AB) + P(AB^c)$ $\qquad$ 3) $P(A^c B^c) = P(A^c) \cdot P(B^c)$

  $\therefore P(AB^c) = P(A) - \underbrace{P(AB)}_{=P(A)P(B)} = P(A)[\underbrace{1 - P(B)]}_{=P(B^c)} = P(A)P(B^c)$

  3) $P(A^c B^c) = P(A \cup B)^c) = 1 - \underbrace{P(A \cup B)}_{\substack{=P(A)+P(B)- P(AB) \\ P(A)P(B)}} = 1 - P(A) - P(B) + P(A)P(B)$

  $\qquad\qquad = [1 - P(A)] - P(B)[1 - P(A)] = [1 - P(A)][1 - P(B)] = P(A^c) \cdot P(B^c)$

- **Conditional Independence** in one sample space $\quad S \qquad P(AB|S) \overset{?}{=} P(A|S) \cdot P(B|S)$

  says nothing about conditional indep. in another

  $\qquad\qquad\qquad\qquad\qquad\qquad\qquad\qquad \hat{S} \qquad P(AB|\hat{S}) \overset{?}{=} P(A|\hat{S}) \cdot P(B|\hat{S})$

Events G and E of the last slide are disjoint so their intersection is null $G \cdot E = \phi$, and accordingly $P(G \cdot E) = 0$. If these two disjoint events were also independent we would require the joint probability to be the product of their individual probabilities, so we would have $0 = P(G \cdot E) = P(G)*P(E)$. But this result cannot be satisfied unless one or both are the null event, so that either $P(G)$ or $P(E)$ is zero. Thus events G and E are not *independent*; in fact they are totally *dependent* since an element in G is specifically excluded from also being in E.

The primary definition of independence for two events A and B states that their joint probability $P(A \cdot B)$ factors into a product $P(A)*P(B)$. There are several other useful forms for independence involving joint distributions composed with A, B and their complements $A^c$, $B^c$; these alternative forms expressing independence along with their formal set theory proofs are given in the slide.

Similarly, *conditional independence* requires that a *joint conditional* probability such as $P(A \cdot B|S)$ factors into the product of conditionals $P(A|S)*P(B|S)$. It is also important to understand that the independence of two events in *one space S* says nothing about their independence in a reduced sample space *S-cap* because of the parameter constraints imposed when forming S-cap.

# Independent Sets of Events & Repeated Trials

## 5.1.2.1 Conditional Independence:  4-Sided Dice (Two Sample Spaces)

### Conditional Independence: 4-Sided Dice  (Two Sample Spaces)

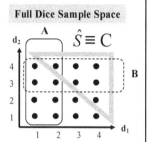

**Sample Space**   $S = \{(d_1, d_2) ; \ d_1, d_2 = 1, 2, 3, 4\}$ ; $P(S) = 1$

**Event A**: $d_1 = 1$ or 2     P(A)= 8/16

**Event B**: $d_2 = 3$ or 4     P(B)= 8/16

**Original Event Space S:**  $\{(d_1, d_2)\}$    *Unrestricted Dice "Values"*

$$\underbrace{P(AB \mid S)}_{4/16} \overset{?}{=} \underbrace{P(A \mid S)}_{8/16} \cdot \underbrace{P(B \mid S)}_{8/16}$$

**YES** Cond. Indep.
**Directly from Figure**

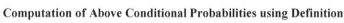

**New Sample Space is Event C**   $\hat{S} \equiv C = \{(d_1, d_2) ; \ Sum = d_1 + d_2 \geq 6\}$ ;  $P(\hat{S}) = 6/16$

$\hat{S} = \{(d_1, d_2) \in C\}$    *Dice Values Restricted to Event C*

$$\underbrace{P(AB \mid \hat{S})}_{1/6} \overset{?}{=} \underbrace{P(A \mid \hat{S})}_{1/6} \cdot \underbrace{P(B \mid \hat{S})}_{5/6}$$

**NOT** Cond. Indep. !!
**Directly from Figure**

**Computation of Above Conditional Probabilities using Definition**

$$P(AB \mid \hat{S}) = \frac{P(AB\hat{S})}{P(\hat{S})} = \frac{1/16}{6/16} = \frac{1}{6} \qquad P(A \mid \hat{S}) = \frac{P(A\hat{S})}{P(\hat{S})} = \frac{1/16}{6/16} = \frac{1}{6} \qquad P(B \mid \hat{S}) = \frac{P(B\hat{S})}{P(\hat{S})} = \frac{5/16}{6/16} = \frac{5}{6}$$

In the $d_1$-$d_2$ sample space for 4-sided dice we define two events A and B (rectangular regions in figure) with probabilities P(A)=8/16, P(B)=8/16, and joint probability P(AB) =4/16.  In the full event space S we test the equality P(AB|S)=P(A|S)∗P(B|S) to find 4/16= (8/16)(8/16) which is satisfied and shows that A and B are independent in S.  Now consider the reduced event space S-cap defined by the constraint $d_1 + d_2 \geq 6$ as shown by the grey triangle and note that there are exactly 6 points in this reduced sample space.   If we now test for independence in S-cap using the conditional joint probability test P(AB|S-cap)=P(A|S-cap)∗P(B|S-cap) we find directly from the figure that the joint conditional probability 1/6 does not equal the product of the conditionals (1/6)∗(5/6) so we conclude that *in S-cap A and B are no longer independent.*  Note that we always have the option to compute these S-cap conditional probabilities using their definitions directly in the original sample space S as P(AB|S-cap)=P(AB·S-cap)/P(S-cap) = (1/16) / (6/16) = 1/6, *etc.* as given on the bottom of the slide.

## 5.1.2.2 Independence Example: Raffle Tickets

---

### Independence Example: Raffle Tickets

- **Ex. 1**: **500 Raffle Tickets; Two Prizes**;       Prob of winning *at least one* prize

*a) Same Ticket allowed to win both prizes*

- Random draw **with replacement**    1) $P(W_1 \cup W_2) = \underbrace{P(W_1)}_{=p} + \underbrace{P(W_2)}_{=p} - \underbrace{P(W_1 W_2)}_{=p^2} = 2p - p^2 = 2(.1) - (.1)^2 = \boxed{0.1900}$

- Customer Purchases 50 tickets
- Independent Draws both with   2) $P(W_1 \cup W_2) = 1 - P((W_1 \cup W_2)^c)$

$$p = P(W_1) = P(W_2) = \frac{50}{500} = .10$$

$$= 1 - \underbrace{P(W_1^c W_2^c)}_{\text{indep. events}} = 1 - P(W_1^c)P(W_2^c) = 1 - (.9)^2 = \boxed{0.1900}$$

---

*b) Each ticket can only win one prize*

   **without replacement**; not indep. draws

**Events:** $W_1$ wins on 1st Drawing , $W_2$ wins on 2nd Drawing

**Total Probability from tree outputs:**

$$P(W_1) = P(W_1 W_2) + P(W_1 W_2^c) = \frac{1}{10} \cdot \frac{49}{499} + \frac{1}{10} \cdot \frac{450}{499} = \frac{1}{10}$$

$$P(W_2) = P(W_2 W_1) + P(W_2 W_1^c) = \frac{1}{10} \cdot \frac{49}{499} + \frac{9}{10} \cdot \frac{50}{499} = \frac{1}{10}$$

$$P(W_1 W_2) = \frac{1}{10} \cdot \frac{49}{499} = .00982 \neq P(W_1) \cdot P(W_2) = .01$$

**Conditional Probability from 2nd branches:**

$$P(W_2 \mid W_1) = \frac{49}{499} = .0982 \neq P(W_2) = 0.1$$

Criteria for <u>independence</u> not satisfied "exactly"

$$P(W_2 \mid W_1^c) = \frac{50}{499} = .1002 \neq P(W_2) = 0.1 \quad \text{"weakly dependent"}$$

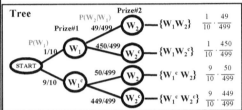

**Tree**

Prize#1    Prize#2

$P(W_1) = 1/10$   $P(W_2|W_1) = 49/499$

$W_1 \to W_2 \to \{W_1 W_2\} \quad \frac{1}{10} \cdot \frac{49}{499}$

$450/499 \to W_2^c \to \{W_1 W_2^c\} \quad \frac{1}{10} \cdot \frac{450}{499}$

START

$9/10 \to W_1^c$

$50/499 \to W_2 \to \{W_1^c W_2\} \quad \frac{9}{10} \cdot \frac{50}{499}$

$449/499 \to W_2^c \to \{W_1^c W_2^c\} \quad \frac{9}{10} \cdot \frac{449}{499}$

1) $P(W_1 \cup W_2) = P(W_1) + P(W_2) - P(W_1 W_2)$

$$= .1 + .1 - \frac{1}{10} \cdot \frac{49}{499} = \boxed{0.1902}$$

2) $P(W_1 \cup W_2) = 1 - P((W_1 \cup W_2)^c) = 1 - P(W_1^c W_2^c)$

$$= 1 - \frac{9}{10} \cdot \frac{449}{499} = \boxed{0.1902}$$

---

In this example 500 raffle tickets are sold for chances to win two prizes. One individual purchases 50 tickets and we are asked the probability of winning "at least one" prize in two draws a) under the condition that the winning ticket of the first draw is placed back into the bowl and b) when it is not. If we let $W_1$ be the event that the individual wins prize #1, and $W_2$ the event that he wins prize#2, then we must calculate $P(W_1 \cup W_2)$.

**In part a) the same ticket is allowed to win both prizes** so we have *two independent drawings* (with replacement). The probability of winning for each draw is $p = P(W_1) = P(W_2) = 50/500 = 1/10$, and we can expand the probability of the union $P(W_1 \cup W_2)$ using either inclusion/exclusion or DeMorgan's laws as follows:

$$P(W_1 \cup W_2) = P(W_1) + P(W_2) - P(W_1 W_2) = p + p - P(W_1)*P(W_2) = 2p - p^2 = .19,$$
$$P(W_1 \cup W_2) = 1 - P[(W_1 \cup W_2)^c] = 1 - P(W_1^c W_2^c) = 1 - P(W_1^c)P(W_2^c) = 1 - .9^2 = .19$$

**In part b) the same ticket is allowed to win only one prize** and then removed from the bowl, so we have *two dependent drawings* (without replacement). Note that because the winning ticket is removed, the individual has either a lesser or greater chance of winning prize#2 depending respectively upon whether or not he won the first prize; thus we have $P(W_2 \mid W_1) = 49/499$ if he won prize#1 and $P(W_2 \mid W_1^c) = 50/499$ if he did not. Perhaps, surprisingly we again have $P(W_1) = P(W_2) = 1/10$ for the two draws, which can be verified by using the law of total probability and picking terms directly off the tree. Tree computations also show that the joint probability $P(W_1 W_2) = (1/10)*(49/499) = .00982$ is not equal $P(W_1)*P(W_2) = .01$ so that the variables are *not independent*. Thus, the same two methods as in part a) yield a *slightly larger probability for the dependent draws*

$$P(W_1 \cup W_2) = P(W_1) + P(W_2) - P(W_1 W_2) = .1 + .1 - P(W_2 \mid W_1)*P(W_1) = .2 - (49/499)*.1 = .1902$$
$$P(W_1 \cup W_2) = 1 - P[(W_1 \cup W_2)^c] = 1 - P(W_1^c W_2^c) = 1 - .9*(449/499) = .1902$$

This result for dependent draws is *only slightly larger* because the probability for two failures here $.9*(449/499) = .8098$ is *slightly smaller* than .81 for case a). We see that, for a large number (500) the difference between case a) and case b) is negligible, so the two events are "nearly independent" or equivalently "weakly dependent." Thus, when describing physical phenomena involving a large number of events, we can often assume independence even though it is not strictly true because the difference is negligible as in this example.

## 5.1.2.3 Independence Example: Communication Channel Redundant Bits

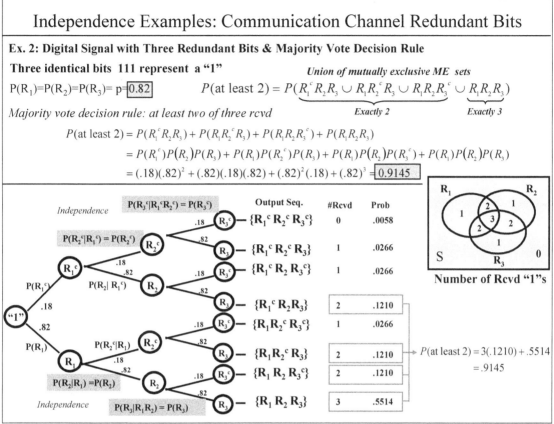

Independence Examples: Communication Channel Redundant Bits

**Ex. 2: Digital Signal with Three Redundant Bits & Majority Vote Decision Rule**

**Three identical bits  111  represent a "1"**

$P(R_1)=P(R_2)=P(R_3)= p=\boxed{0.82}$

*Union of mutually exclusive ME  sets*

$$P(\text{at least 2}) = P(\underbrace{R_1^c R_2 R_3 \cup R_1 R_2^c R_3 \cup R_1 R_2 R_3^c}_{\text{Exactly 2}} \cup \underbrace{R_1 R_2 R_3}_{\text{Exactly 3}})$$

*Majority vote decision rule: at least two of three rcvd*

$$P(\text{at least 2}) = P(R_1^c R_2 R_3) + P(R_1 R_2^c R_3) + P(R_1 R_2 R_3^c) + P(R_1 R_2 R_3)$$

$$= P(R_1^c)P(R_2)P(R_3) + P(R_1)P(R_2^c)P(R_3) + P(R_1)P(R_2)P(R_3^c) + P(R_1)P(R_2)P(R_3)$$

$$= (.18)(.82)^2 + (.82)(.18)(.82) + (.82)^2(.18) + (.82)^3 = \boxed{0.9145}$$

The communication example in this slide uses the redundant sequence "111" to represent a digital "1" and invokes a *majority decision rule* that requires *at least two of three* "1"s to be received correctly in order to declare a "1". This triple redundancy creates "overhead" in the data stream; however it allows for some "diversity" because a *detection is declared* even if one of the three bits drops out of the signal because of bad geometry or interference.

In the *upper panel* we assume that each of the independent bits has a common probability of detection of p = 0.82, say, and invoke the majority decision criterion by taking the union of the mutually exclusive ME events {R₁ᶜR₂R₃, R₁R₂ᶜR₃, R₁R₂R₃ᶜ, R₁R₂R₃} corresponding to the regions labeled with 2 and 3 correctly received bits in the Venn diagram. The calculation is straight-forward provided the concepts of *independence* and *mutual exclusivity* are employed to expand the probability $P(R_1^cR_2R_3 \cup R_1R_2^cR_3 \cup R_1R_2R_3^c \cup R_1R_2R_3)$. Specifically, ME allows us to compute the probability of this union as the sum of individual event probabilities of the four terms, while independence allows us to write each of these terms as a product, *e.g.*, $P(R_1^cR_2R_3) = P(R_1^c)P(R_2)P(R_3)=(.18)(.82)(.82)$.

The *bottom panel* displays the corresponding tree diagram for majority detection of a "1" by measuring three independent bits and employing the majority decision rule which declares a "1" whenever two or three bits are received correctly ("at least two bits"). The explicit values "Output Seq, #Rcvd, and Prob" are displayed and the relevant boxed outputs are summed to give P(at least 2) = .9145. Note that the tree outputs are mutually exclusive because they are labeled by a sequence of distinct nodes; also the independence of the measurements allows us to write $P(R_2|R_1)=P(R_2)=.82$ and $P(R_3|R_1 R_2) =P(R_3)=.82$, in the branch labels.

# Independent Sets of Events & Repeated Trials

## 5.1.3 Mutual Independence - 3 or More Events

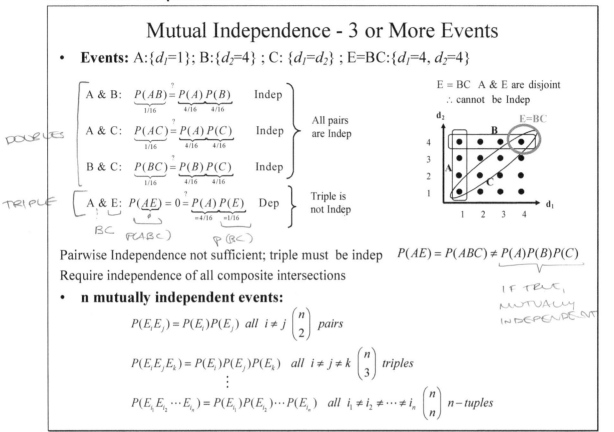

Pairwise Independence not sufficient; triple must be indep  $P(AE) = P(ABC) \neq P(A)P(B)P(C)$

Require independence of all composite intersections

- **n mutually independent events:**

$$P(E_i E_j) = P(E_i)P(E_j) \quad all \ i \neq j \quad \binom{n}{2} \ pairs$$

$$P(E_i E_j E_k) = P(E_i)P(E_j)P(E_k) \quad all \ i \neq j \neq k \quad \binom{n}{3} \ triples$$

$$\vdots$$

$$P(E_{i_1} E_{i_2} \cdots E_{i_n}) = P(E_{i_1})P(E_{i_2}) \cdots P(E_{i_n}) \quad all \ i_1 \neq i_2 \neq \cdots \neq i_n \quad \binom{n}{n} \ n-tuples$$

A set of mutually independent events is one for which all joint probabilities on event pairs, on event triples, ..., on event n-tuples are also be independent and thus factor into products of the individual event probabilities. In particular for three events A, B, and C shown on the slide this requires the pairs AB, AC, and BC and the triple ABC all be independent. For a pair of 4-sided dice, the joint and individual probabilities for the three defined events A,B,C are easily read off the figure; it is then easy to verify that the joint probabilities factor and hence the events are *pair-wise independent*; however, we must also show that P(ABC)= P(A)∗P(B)∗P(C). To do this, we define the event E=BC and test the "equivalent pair" P(AE) to see if it equals P(A)∗P(E); this actually tests the triple since P(A)∗P(E) = P(A)∗P(BC)= P(A)∗[P(B)∗P(C)] which is factorization of the triple. Inspection of the figure shows that A and E are disjoint so they cannot be independent and hence the triple does not factor and accordingly, the set {A,B,C} is not mutually independent *even though the elements are pair-wise independent*.

The generalization to a **set of n mutually independent events** requires independence between all pairs, triples,..., n-tuples, so that every possible composition of these elements is independent. This is a very stringent set of conditions and may be difficult to show for higher dimensions.

## 5.1.4 More Ideas and Examples on Independence

# More Ideas and Examples on Independence

- **Ex.: Three Mutually Independent Events A, B, C**

  Consider composite event E = A∪B

  Show E & C are mutually independent

  $P(EC) = P((A \cup B)C) = P(AC \cup BC)$

  $$= \underbrace{P(AC)}_{P(A)P(C)} + \underbrace{P(BC)}_{P(B)P(C)} - \underbrace{P(AC \cdot BC)}_{P(A)P(B)P(C)}$$

  $$= P(C)[P(A) + P(B) - P(A)P(B)]$$

  $$\boxed{P(EC) = P(C)P(A \cup B) = P(C)P(E)}$$

- **Ex.: Three Mutually Independent Elevators** (Different Motors, Cables, *etc.*)

  Events $E_1$, $E_2$, $E_3$ that each elevator is functioning   $P(E_1) = .7;\ P(E_2) = .8;\ P(E_3) = .9$

  **a) Find Prob exactly one not functioning**   $P(A) = P(E_1^c E_2 E_3) + P(E_1 E_2^c E_3) + P(E_1 E_2 E_3^c)$

  Define Event : $A \equiv E_1^c E_2 E_3 \cup E_1 E_2^c E_3 \cup E_1 E_2 E_3^c$

  $= P(E_1^c)P(E_2)P(E_3) + P(E_1)P(E_2^c)P(E_3) + P(E_1)P(E_2)P(E_3^c)$

  $= (1 - .7)(.8)(.9) + (.7)(1 - .8)(.9) + (.7)(.8)(1 - .9) = .398$

  **b)** Given Event A: "exactly one" not working

  **Find Prob it is Elevator #1, *i.e.*, Event $E_1^c$**

  $P(E_1^c \mid A) = \dfrac{P(E_1^c A)}{P(A)} \neq \dfrac{P(E_1^c)P(A)}{P(A)} = P(E_1^c) = .3$   Because $E_1^c$ and A are not independent

  Event : $E_1^c A \equiv \underbrace{E_1^c E_1^c E_2 E_3}_{E_1^c} \cup \underbrace{E_1^c E_1 E_2^c E_3}_{\varnothing} \cup \underbrace{E_1^c E_1 E_2 E_3^c}_{\varnothing} = E_1^c E_2 E_3$

  $P(E_1^c \mid A) = \dfrac{P(E_1^c A)}{P(A)} = \dfrac{P(E_1^c E_2 E_3)}{P(A)} = \dfrac{P(E_1^c)P(E_2)P(E_3)}{P(A)} = \dfrac{(1-.7)(.8)(.9)}{.398} = .543$

  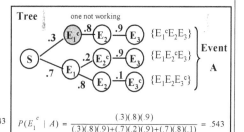

  $P(E_1^c \mid A) = \dfrac{(.3)(.8)(.9)}{(.3)(.8)(.9)+(.7)(.2)(.9)+(.7)(.8)(.1)} = .543$

The slide explores events composed from a set of three mutually independent events. Sometimes the composed event E = A U B is independent of remaining member C of the independent set {A, B, C} (**top panel**). The **lower panels** investigate the working status of a set of three mutually independent elevators {$E_1,E_2,E_3$} and considers an event A defining *exactly one* elevator not working. It turns out that this composite event A is no longer independent of the set {$E_1$, $E_2$, $E_3$}. The three elevators have different motors, cables, computers, *etc.*, and hence are mutually independent. We assume that they operate reliably with probabilities $P(E_1)$=.7, $P(E_2)$=.8, and $P(E_3)$=.9. In part a) below we compute compute the event A that exactly one elevator is working and in part b) we determine the probability that a particular elevator is not working.

**a)** The event A, that exactly one elevator is not working, is easily written down using the following set composition $A = E_1^c E_2 E_3 \cup E_1 E_2^c E_3 \cup E_1 E_2 E_3^c$. The probability of the event A can be computed simply as the sum of the individual probabilities since the three composite events are disjoint (ME). The mutual independence of the events allows us to write the joint probabilities as the products of the individual probabilities and we find P(A) =.398 as shown in the slide.

**b)** In the *new sample space* $A = E_1^c E_2 E_3 \cup E_1 E_2^c E_3 \cup E_1 E_2 E_3^c$ defined by the set of outcomes "exactly one elevator is not working" find the inverse probability that a particular elevator, say elevator #1 (see tree), is not working $E_1^c$; that is we seek the probability $P(E_1^c|A)$. Using the definition of conditional probability we have $P(E_1^c|A) = P(A,E_1^c)/P(A)$ and we would be "tempted" to think A and $E_1^c$ are independent and accordingly, write the joint probability as a product which yields $P(E_1^c|A) = P(A)P(E_1^c)/P(E_1^c)$ =P(A)=.3. However, looking at the joint event $E_1^c A$, we produce three terms, two of which are null because they contain $E_1$, and we find $E_1^c A = E_1^c E_2 E_3$. Thus, although the individual elements {$E_1$, $E_2$, $E_3$} are mutually independent, the construction of A from the E's leaves it no longer independent of $E_1^c$ and as a result we cannot factor the joint probability. Evaluation of $P(E_1^c| A) = P(A,E_1^c) / P(A)$ yields instead $P(E_1^c E_2 E_3)/P(A) = P(E_1^c)*P(E_2)*P(E_3)/P(A)$ =.543.

## 5.1.5 System Reliability Analysis

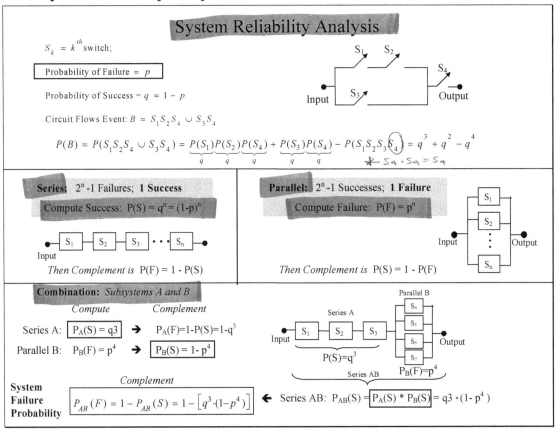

For systems consisting of multiple components the system reliability is assessed by determining all possible paths from the **input** to the **output** and then dividing that number into the *number of paths that fail* at some point along the path. A system consisting of switches which have two positions on and off is a good method to model complex systems.

Traditionally, in system reliability analysis, the **probability of failure is p,** the probability of **success is q=1-p,** and a *closed state* for switch n is denoted by $S_n$. For the system of switches shown in the top panel, current flows through the circuit if "at least one" of the two paths is closed. The probability of success may be written in terms of the union of the two success paths which is the event $B = S_1S_2S_4 \cup S_3S_4$, so $P(B) = P(S_1S_2S_4 \cup S_3S_4) = P(S_1S_2S_4) + P(S_3S_4) - P((S_1S_2S_4) \cdot (S_3S_4))$. If we further assume that all switches are independent elements, each joint probability becomes a product of the individual switch *success probabilities* q. However, care must be taken to evaluate the implied set intersections and replace $S_4 \cdot S_4$ by $S_4$ so as not to square the success probability associated with this component.

This path counting method can be used for any system, but can become tedious for very complex systems. We can simplify things, however, if we recognize that switches in series only succeed if they are all closed so compute $P_{series}(S) = q^n = (1-p)^n$, while switches in parallel, only fail if they are all open so compute $P_{parallel}(F) = p^n$. We may find the complement or each if needed.

Applying this to the system of switches in the bottom panel, we compute

success for 3 switches in series A:    $P_A(S) = q^3$    or    $P_A(F) = 1 - q^3$

failure for 4 switches in parallel B:    $P_B(F) = p^4$    or    $P_B(S) = 1 - p^4$

Combine A and B in series: $P_{A,B \, series}(S) = P_A(S) * P_B(S) = q^3 * (1-p^4)$ or $P_{A,B \, series}(F) = 1 - [q^3 * (1-p^4)]$.

## 5.2 Repeated Independent Bernoulli Trials Yields Binomial

# Repeated Independent Bernoulli Trials Yields Binomial

**Independent "Sub experiments"**
Sub Experiments Diff. Copies of Same Sample space
**i) Flip Coin Repeatedly**: {S: (0,1)} (Bernoulli trials)
**ii) Throw Pair Dice Repeatedly**: {S: $(d_1,d_2)$}
 "Craps" Sum=$(d_1+d_2)$={2,3,4,5,6,7,8,9,10,11,12}

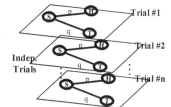

**Ex. Repeated Trials: 3 coin flips**

S(Bernoulli Trial):  $P(H) = p$ ;  $P(T) = q = (1-p)$

Find Prob of 0, 1,2, 3 Heads

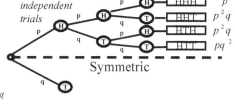

$^3C_0 \qquad P(0H) = q^3$

$^3C_1 \qquad P(1H) = P(HTT) + P(THT) + P(TTH) = pq^2 + qpq + q^2p = 3pq^2$

$^3C_2 \qquad P(2H) = P(HHT) + P(THH) + P(HTH) = p^2q + qp^2 + pqp = 3p^2q$

$^3C_3 \qquad P(3H) = P(HHH) = p^3 \qquad (\cdots + \text{Symmetric Terms})$

$Sum = (p+q)^3 = 1^3 = 1$

**n Coin Flips → Binomial Distribution "b"**

$P(\text{at least one Head}) = 1 - P(0H) = 1 - q^n$

$$P(k - Heads) \equiv b(k;n,p)$$
$$= P_{n,p}(k) = \begin{bmatrix} n \\ k, n-k \end{bmatrix} p^k q^{n-k} = {}^nC_k p^k q^{n-k}$$

The concept of independent trials is a very important one for understanding many physical phenomena. As discussed in the Raffle Ticket example, even if the trials are not strictly independent, the dependence may be so weak that results are essentially the same as if they actually were. To formalize this concept of independence, we envision trials taking place in identical copies of the same sample space that have no way of interacting with each other. Thus, flipping a coin might have copies that look as shown in the top figure with trial#1 to trial #n taking place in separate but identical sample spaces.

Flipping a coin is an experiment with only two outcomes {0,1} or {H,T} and is called a Bernoulli trial. If we flip a coin 3 times we generate a new sample space corresponding to the $2^3 = 8$ possible outcomes 4 of which are shown in the upper tree; the lower half is identical if we swap H<=>T. Note that the binomial expansion $(p+q)^3$ generates the probability for the 4 distinct groupings {3H,0T}, {2H,1T}, {1H,2T}, {0H,3T} in the output of the full tree. Since p+q=1, the sum of the binomial terms generated by $(p+q)^3$ must also be unity and we have

$$^3C_0\, p^0q^3 + {}^3C_1\, p^1q^2 + {}^3C_2\, p^2q^1 + {}^3C_3\, p^3q^0 = 1$$

This set of terms defines the binomial distribution

$$b(k;3,p) = {}^3C_k\, p^kq^{3-k} \text{ for k=0,1,2,3.}$$

The general expression for the $k^{th}$ term of the $n^{th}$ order binomial distribution is

$$b(k;n,p) = {}^nC_k\, p^kq^{n-k} \quad \text{for k=0,1,2, ..., n}$$

where n= #trials, k=#succ (Heads) and p= *single trial probability of success.*

# Independent Sets of Events & Repeated Trials

## 5.2.1 Repeated Independent Trials: Airline Overbooking

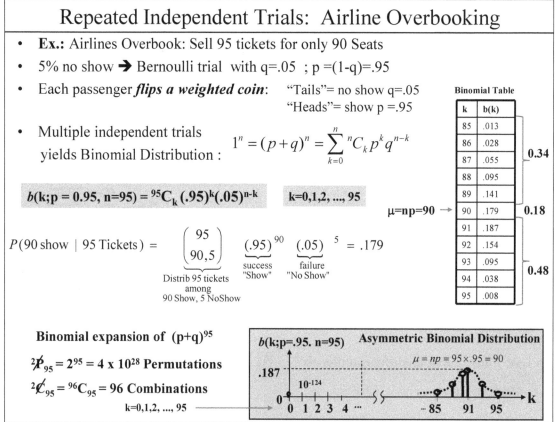

Airlines often overbook, for example, by selling 95 tickets for 90 available seats; they bank on a 5% no show rate which is modeled as a Bernoulli trial (coin flip) with no show rate $q=.05$ and $p=1-q=.95$. So for $n=95$ tickets we want the probability that *exactly* $k=90$ seats will actually be filled and this is given by the binomial term

$$b(90; 95,.05) = P(90\ \text{show}|95\ \text{tickets}) = {}^{95}C_{90}(.95)^{90}(.05)^{5} = .179$$

A more complete analysis requires all terms in the asymmetric binomial distribution given in the table and illustrated by the plot at the bottom of the slide; the value at $k=0$ corresponds to none of the 95 passengers showing up and is symbolized by a "dot" at the origin with magnitude $2.5 \times 10^{-124}$. The table shows that for $k = 93$, the probability is .095 and for $k = 95$ (all show) the probability is extremely small, .008; the peak value .187 occurs at $k= 91$ passengers. Moreover, the average number of passengers is $\mu = np = 95(.95) = 90.25$ which equals the number of available seats. Note also that the distribution is reasonably symmetric about the mean at "90" and the significant contributions are from $k = 85$ to 95. The table shows that the probability of under-booking ($<90$) is $P(k < 90) = .34$ while that for overbooking ($>90$) is $P(k > 90) = .48$; the probability of exactly filling the aircraft is $P(k=90) = .18$. Thus, the trade-off is that 34% of the time 5 unfilled seats with "retail revenues of say $200" are lost, yielding $5*\$200*.34 = \$340$ *versus* 48% of the time 5 "wholesale revenues of say $100" are lost (giving away flights) $5*\$100*.48 = \$240$. Clearly over-booking wins by maximizing occupancy and minimizing loses! Some passengers also win by getting free tickets!

It is interesting to note that a tree diagram results from 95 random draws from "show" and "no show". This yields $2^{95} = 4 \times 10^{28}$ final states or permutations (with replacement), but only "slash"-${}^{2}C_{95} = {}^{97-1}C_{95} = 96$ combinations (w/replacement) corresponding to the distinct terms of the binomial distribution for $k = 0, \ldots, 95$.

## 5.2.2 Repeated Trials: "Craps After Point is Established"

# Repeated Trials: "Craps After Point is Established"

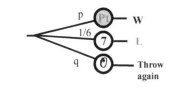

$p$ = prob of rolling a "pt"

1/6 = prob of rolling a "7"

$q$ = Prob of rolling "Other"

$$p + q + \frac{1}{6} = 1$$

**Dice Probability Table**

| $S = d_1 + d_2$ | #Ways | #Prob |
|---|---|---|
| 2, 12 | 1 | 1/36 |
| 3, 11 | 2 | 2/36 |
| 4, 10 | 3 | 3/36 |
| 5, 9 | 4 | 4/36 |
| 6, 8 | 5 | 5/36 |
| 7 | 6 | 6/36 |

Points { 4, 10 ; 5, 9 ; 6, 8

$$P(win \mid pt) = p + qp + q^2 p + q^3 p + \cdots = p(1 + q + q^2 + \cdots)$$

$$= \frac{p}{1-q} = \frac{p}{p+1/6}$$

Here is another look at the game of craps, using the idea of repeated trials to develop a general formula for the probability of winning after the point has been established. Given the "point," we define the conditional probabilities emanating from the "other" (o) node as (i) p, the probability of rolling the point and winning, (ii) 1/6, the probability of rolling a "7" and losing, and (iii) q, the probability of taking another throw (other). These three probabilities exhaust all possibilities emanating from the "o" node and therefore must sum to unity: p+q+1/6 = 1.

Thus, drawing a tree starting at "o" on the second throw, we have the three branches whose probabilities we have just defined and this sub-tree repeats itself an infinite number of times. Since each trial is independent with the same p (win) and q (throw again) there are an infinite number of ways to win namely win on the 1st throw with probability "p" or win on the 2nd throw with probability "qp", or on the 3rd throw "q²p" , on the 4th throw "q³p", ... , so we have the infinite geometric sum
$$P(win \mid pt) = q + qp + q^2 p + q^3 p + \ldots = p / (1-q).$$
Replacing (1-q) in this expression with (p+1/6) yields
$$P(win \mid pt) = p/(p+1/6),$$
which is the general formula for the probability of winning "given the point," where p is the appropriate value taken from the Dice Probability Table.

## 5.2.3 Gambler's Ruin

# Gambler's Ruin

- **Difference Equations** –
  - Local Conditions on Probability
  - Yields recursion equation
  - $P(A) = p$ ;  $P(B) = q$
  - **A starts with "k-units"**
  - **B starts with "(n-k)-units"**
  - Wager 1 unit until ...
  - one player is "ruined" (out of money)

  $P_k$ = prob A eventually wins starting with $k$ units

  $P_{k-1}$ = prob A eventually wins starting with $(k-1)$ units

  $P_{k+1}$ = prob A eventually wins starting with $(k+1)$ units

- **Recursion Equation:** Prob A wins starting with k ($\mathbf{P_k}$) is related to Prob A wins starting with k-1 ($\mathbf{P_{k-1}}$) & A wins starting with k+1 ($\mathbf{P_{k+1}}$)

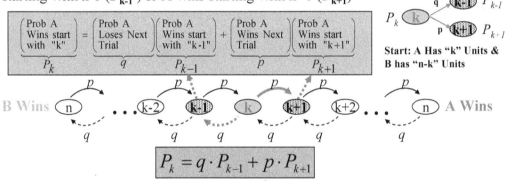

$$P_k = q \cdot P_{k-1} + p \cdot P_{k+1}$$

The Gambler's Ruin is a classic problem using repeated trials and local conditions on probability to construct a recursion equation that is solved by elementary methods. Consider two players A and B with probability of winning $P(A) = p$ and $P(B) = q = 1$-p and a fixed amount "n units" of money for betting. Assume that A starts with "k units" and B starts with "(n-k) units" and further that they only bet 1 unit each time until one player has 0 units ("ruined"). The problem is to determine the probability that *A eventually wins if he starts with k units*; letting $E_k$ denote this event, then $P_k = P(E_k)$ is the probability that we seek. This starting event can be followed by a loss with probability q which leads to state k-1 or a win with probability p which leads to state k+1 as illustrated to the right of the large boxed equation in the bottom panel. This basic choice results in the chain of events below.

The "word equation" is stated in the grey box (bottom panel), and the graphic immediately below it shows the states corresponding to player A having k-1, k, k+1 units and the curved arrows show transitions between the various neighboring states. If A is at state k (grey state), there are only two (ME and CE) outcomes *for his next turn*, namely, (i) he can lose with probability q and wind up in state "k-1" (broken red arrow from k to k-1) or (ii) he can win with probability p and wind up in state "k+1"(solid red arrow from k to k+1). Thus, if A is to *eventually win starting with k units*, he *must also win after transitioning* with probabilities p or q to one of these two neighboring states; hence $P(E_k)$ must equal the weighted sum $q*P(E_{k-1}) + p*P(E_{k+1})$, thus establishing the second order recursion relation:  $P_k = qP_{k-1} + pP_{k+1}$. We solve this second order difference equation on the next slide.

## 5.2.3.1 Gambler's Ruin:  Solution to Difference Equation

# Gambler's Ruin:  Solution to Difference Equation

| Difference Equation | Boundary Conditions |
|---|---|
| $$P_k = q \cdot P_{k-1} + p \cdot P_{k+1}$$ | *Start with "A wins" i.e., has all n points k=n* $\quad P_k\big|_{k=n} = 1$ <br> *Start with "A loses" i.e., has all no points k=0* $\quad P_k\big|_{k=0} = 0$ |

**Trial Soln:** $\quad P_k = \lambda^k \Rightarrow \lambda^k - p\lambda^{k+1} - q\lambda^{k-1} = \lambda^{k-1}(\lambda^1 - p\lambda^2 - q) = 0$

$$\lambda = \frac{1 \pm \sqrt{1-4pq}}{2p} = \frac{1 \pm \sqrt{1-4(1-q)q}}{2p} = \frac{1 \pm \sqrt{(2q-1)^2}}{2p} = \frac{1 \pm (2q-1)}{2p} = \begin{cases} q/p \\ 1 \end{cases} \quad \textbf{Two Roots}$$

**Case A)** $q \neq p$ two distinct roots 1, q/p $\quad P_k = C_1 (q/p)^k + C_2 (1)^k \quad \rightarrow \quad P_k = -\dfrac{1}{1-(q/p)^n} \cdot (q/p)^k + \dfrac{1}{1-(q/p)^n} \quad\boxed{ P_k = \dfrac{1-(q/p)^k}{1-(q/p)^n}}$

**Apply BC** $\quad \left.\begin{array}{l} P_0 = C_1 + C_2 = 0 \\ P_n = C_1 (q/p)^n + C_2 \end{array}\right\} \Rightarrow C_1 = -C_2 = -\dfrac{1}{1-(q/p)^n}$

**Let q/p = 1 − ε & take lim$_{ε\to0}$ yields Case B)** $\quad P_k(q/p=1) = \underset{\varepsilon \to 0}{Lim}\left(\dfrac{1-(1-\varepsilon)^k}{1-(1-\varepsilon)^n}\right) = \underset{\varepsilon \to 0}{Lim}\left(\dfrac{1-(1-k\varepsilon)}{1-(1-n\varepsilon)}\right) = \underset{\varepsilon \to 0}{Lim}\dfrac{k\varepsilon}{n\varepsilon} = \dfrac{k}{n} \quad \boxed{P_k(q/p=1) = \dfrac{k}{n}}$

**Case B)** q /p =1 $\lambda = 1$ a double root $\quad P_k = C_1 (1)^k + C_2 \cdot k \cdot (1)^k$

**Apply BC** $\quad \left.\begin{array}{l} P_0 = C_1 + C_2(0) = 0 \\ P_n = C_1 + C_2(n) = 1 \end{array}\right\} \Rightarrow C_1 = 0 \; ; \; C_2 = \dfrac{1}{n}$ $\qquad$ For a "fair game" p=q=1/2 A's probability of winning increases linearly with the initial "stake" of k-units. $\quad \boxed{P_k(q/p=1) = \dfrac{k}{n}}$

**Case C)** q /p ≪1 $\qquad$ k=1, $p_k$ = 80%; $\quad$ k=2, pk = 90%; $\quad$ k=3, pk = 99% $\qquad \boxed{p_k = 1-(q/p)^k}$

The recursion relationship is easily solved as a second order difference equation by using a trial solution of the form $P_k = \lambda^k$ (analogous to $e^{\lambda t}$ used for ordinary differential equations). Substitution of this trial solution into the recursion for $P_k$, yields a quadratic equation in $\lambda$ with two solutions $\lambda_1 = q/p$ and $\lambda_2 = 1$. For the case $q \neq p$, the general solution is written as a linear combination of the two as

$$P_k = C_1 (\lambda_1)^k + C_2 (\lambda_2)^k = C_1 (p/q)^k + C_2 (1)^k.$$

There are two boundary conditions stating the following extremes:
(i) If A starts with "0 units" he has lost before he even begins, so his probability of winning is $P_k\big|_{k=0} = 0$, and
(ii) If A starts with "n units" he has already won, so his probability of winning is certain, *i.e.*, $P_k\big|_{k=n} = 1$
These boundary conditions determine $C_1$ and $C_2$ and hence the solution given in the boxed equation

$$P_k = [1-(q/p)^k] / [1-(q/p)^n]$$

The case  p=q can be obtained by taking the limit of the above as $q \to p$ or by direct solution of the difference equation for a double root as given in case B) below; in either case the result is

$$P_k = k/n$$

These two results show that with equal odds (when p=q ) "A"s probability of winning is directly proportional to k the number of units he starts with; that is, a fair game favors the one who starts with the most units k.
At the other extreme if B has a small probability of winning q<< p, an approximate solution is

$$P_k \sim [1-(q/p)^k] * [1+(q/p)^n] = 1-(q/p)^k +(q/p)^n = 1- (q/p)^k \{1-(q/p)^{n-k} \}$$

So with a large win advantage for A, even if he starts with very few units k<<n, he still has a large probability of winning. Further, we can neglect the last term to approximate $P_k \sim 1- (q/p)^k$. For example, taking q/p=.2, even if "A" starts out with just k=1 unit, his probability of winning $P_k \sim 1- (.2)^1 = 80\%$; with k=2 this increases to 96%; k=3 yields 99%.

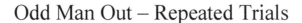

## 5.2.3.2 Odd Man Out – Repeated Trials

---

# Odd Man Out – Repeated Trials

**Game of "Odd Man Out":** *Coin is flipped repeatedly until one is different from all others thereby breaking the "run" of heads  HHH"T"  or tails  TTTT"H"*
**Context:** *Three or more people sitting around a table; "odd man out" pays for meal*

Game for n = 3 with Fair Coin : $\left.\begin{array}{c} P(H)=.5 \\ P(T)=.5 \end{array}\right\}$  $\underset{\substack{\text{no}\\\text{odd}\\\text{man}}}{HHH},\quad \underset{\substack{\text{yes}\\\text{odd}\\\text{man}}}{H\,"T"\,H},\quad \underset{\substack{\text{yes}\\\text{odd}\\\text{man}}}{TT\,"H"}\quad \cdots\quad \underset{\substack{\text{no}\\\text{odd}\\\text{man}}}{TTT}$

$2^3$ = 8 equally likely outcomes  # ways to get one different $^3C_1$ = 3 and $^3C_2$ = 3

$\therefore\ P(out) = \dfrac{3+3}{8} = \dfrac{3}{4}$

**Odd-man-out for p=q=.5**

| n | P ("Out") |
|---|---|
| 3 | 0.75 |
| 4 | 0.50 |
| 5 | 0.31 |
| 10 | 0.020 |
| 100 | $1.58 \times 10^{-28}$ |

Game for n with "unfair" Coin: $\left.\begin{array}{c} P(H)=p \\ P(T)=q=1-p \end{array}\right\}$

$P(out) = P("1H"\ in\ "n\text{-}Thrws"\ ) + P("1T"\ in\ "n\text{-}Thrws"\ )$

$P(out) = \underbrace{^nC_1}_{\substack{\text{arrangements}}}\ \underbrace{p^1 q^{n-1}}_{\text{coin flips}} + {}^nC_{n-1}\ \underbrace{p^{n-1}q^1}_{\text{coin flips}} = npq^{n-1} + np^{n-1}q$

Check:  n = 3,  p = q = 1 / 2   $P(out) = 3\dfrac{1}{2}\left(\dfrac{1}{2}\right)^{3-1} + 3\left(\dfrac{1}{2}\right)^{3-1}\dfrac{1}{2} = \dfrac{3}{4}$

---

"Odd man out"  is one way of deciding who pays for lunch when there are three or more people in a lunch group.  A coin is flipped in turn by each person until one player's coin is different from all the others and he or she is declared "odd man out" and pays for lunch.  Clearly, this does not work with only two people since they are either the same or different and in either case there is no distinction; however, adding a third person allows for an odd man out.  The simple case of three people flipping a fair coin gives $2^3$ = 8 possible outcomes and to get one different from the others we can either place 1"H" among 2"T" in $^3C_1$ =3 ways  or place 1"T" among 2 "H" in $^3C_2$ ways.  Thus the probability of obtaining an odd man out is

$$P("out") = [^3C_1 + {}^3C_2] / 2^3 = .75.$$

For n people, this generalizes to

$$P("out") = [^nC_1 + {}^nC_{n-1}] / 2^n = 2n/ 2^n$$

The table shows that the probability of getting this "lunch paying scheme" to work decreases rapidly with increasing n=3, 4, 5, 10, 100.  While for n=3, the probability of having an odd man out is 75%, it is only 50% for n=4, only 2% at n=10, and it is virtually impossible for n=100.

Now one might think to bring an unfair coin (p ≠ q) with P(H)=p and P(T)=q=1-p.  The solution now requires the sum of two terms generated by the binomial $(p+q)^n$ corresponding to {1"H", (n-1)"T"} and {(n-1) "H", 1"T"}, and yields the following general result for n people

$$P("out") = {}^nC_1 p^1 q^{n-1} + {}^nC_{n-1}\ p^{n-1}q^1 = n \cdot [pq^{n-1} + p^{n-1}q]$$

## 5.2.4 Repeated Independent Trials with 2 and 3 Outcomes

# Repeated Independent Trials with 2 and 3 Outcomes

| | Binomial (2) | Trinomial (3) |
|---|---|---|
| Sample Space | $S=\{H,T\}$    "**2-outcomes**" | $S=\{T,F,N\}$    "**3-outcomes**" |
| Prob(1 trial) | $P(H)=p, P(T)=q$ | $P(T)=p, \; P(F)=q \, , \; P(N)=r$ |
| $\sum_{Outcomes} P(1-trial)$ | $p+q=1$ <br> " 2-outcomes" | $p+q+r=1$ <br> " 3-outcomes" |
| n-indep Trials | $S^n = \underset{n-factors}{\underline{S \times S \times \cdots \times S}}$ | $S^n = \underset{n-factors}{\underline{S \times S \times \cdots \times S}}$ |
| "One" Sequence <br><br> Out of $2^n$ or $3^n$ Seq. | $S^n = \underset{n-terms}{\underline{\{HTTH\cdots TH\}}}$   $^2\mathcal{P}_n = 2^n$ Seq <br> $k-"H" \;\; (n-k)-"T"$ <br> $k+(n-k)=n$ | $S^n = \underset{n-terms}{\underline{\{TTFNN\cdots NTF\}}}$   $^3\mathcal{P}_n = 3^n$ Seq <br> $k_1-"T" \;\; k_2-"F" \;\; k_3-"N"$ <br> $k_1+k_2+k_3=n$ |
| **Prob of "One" Seq.** <br> k-Heads and n-k Tails | $p^k q^{n-k}$ | $p^{k_1} q^{k_2} r^{k_3}$ |
| **Prob of "All Seq."** <br> k-Heads and n-k Tails | $\binom{n}{k,(n-k)} p^k q^{n-k} = \dfrac{n!}{k!(n-k)!} p^k q^{n-k}$ | $\binom{n}{k_1,k_2,k_3} p^{k_1} q^{k_2} r^{k_3} = \dfrac{n!}{k_1!k_2!k_3!} p^{k_1} q^{k_2} r^{k_3}$ |
| Ordered Subsets | $2$,    Sizes $k$ and $n-k$ | $3$,    Sizes: $k_1, k_2, k_3$ |
| **Algebraic Generator** <br><br> **Normalization** <br> **to unity** | Sum of $^2\mathcal{C}_n =$ (n+1) terms   $1^n = (p+q)^n = \sum_{k=0}^{n}\binom{n}{k,n-k} p^k q^{n-k}$ | Sum of $^3\mathcal{C}_n =$ (n+1)(n+2)/2 terms   $1^n = (p+q+r)^n = \sum_{k_1+k_2+k_3=n}\binom{n}{k_1,k_2,k_3} p^{k_1} q^{k_2} r^{k_3}$ |

A single Bernoulli trial results in two outcomes {success =1, failure=0} with probability of success p and failure q=1-p. We have seen how n independent Bernoulli trials generates a new sample space representing the number of successes k in n trials, for a *single trial probability of success* p; each member of this new sample space is expressed as a binomial term $b(k; n, p) = {}^nC_k \, p^k q^{n-k}$. In an analogous fashion, n independent trials in a sample space with 3 outcomes {True, False, Neither} with single trial probabilities of success {p, q, r} generates a trinomial sample space. The general case of n independent trials in a sample space with r distinct outcomes {$s_1, s_2, \dots s_r$} and with single trial probabilities of success{ $p_1, p_2, \dots, p_r$} generates a multinomial sample space. The table summarizes characteristics of the binomial and trinomial sample spaces in terms of (i) the underlying sample space with 2 or 3 outcomes, (ii) the structure of a single sequence, (iii) the multiplicity coefficient for each sequence, (iv) the algebraic generator for all terms of the expansion, and (iv) the normalization to unity that makes it a PMF.

As an example of a multinomial we have, a pair 4-sided dice realized by n=2 independent trials in a sample space with 4 outcomes(faces) and single trial probability of success {$p_1, p_2, p_3, p_4$}; this generates the following multinomial distribution containing 10 distinct groupings as shown below

$$(p_1+p_2+p_3+p_4)^2 = {}^2C_{2000} \, p_1^2 p_2^0 p_3^0 p_4^0 + {}^2C_{0200} \, p_1^0 p_2^2 p_3^0 p_4^0 + {}^2C_{0020} \, p_1^0 p_2^0 p_3^2 p_4^0 + {}^2C_{0002} \, p_1^0 p_2^0 p_3^0 p_4^2$$
$$+ {}^2C_{1100} \, p_1^1 p_2^1 p_3^0 p_4^0 + {}^2C_{1010} \, p_1^1 p_2^0 p_3^1 p_4^0 + {}^2C_{1001} \, p_1^1 p_2^0 p_3^0 p_4^1$$
$$+ {}^2C_{0110} \, p_1^0 p_2^1 p_3^1 p_4^0 + {}^2C_{0101} \, p_1^0 p_2^1 p_3^0 p_4^1 + {}^2C_{0011} \, p_1^0 p_2^0 p_3^1 p_4^1$$

$^4\mathcal{C}_2 = {}^5C_2 = 10 \text{ terms}$

The above terms can be grouped by their dice sums {2,3,4,5,6,7,8} to yield the outcome probabilities for this unfair set of dice. If we set $p_1 = p_2 = p_3 = p_4 = 1/4$ the above results are equivalent to those generated by the "dice generating polynomial" $(x^1+x^2+x^3+x^4)^2 /16 = [\; \underline{1}x^2+\underline{2}x^3+\underline{3}x^4 +\underline{4}x^5+\underline{3}x^6+\underline{2}x^7+\underline{1}x^8]/16$, where the exponents are the dice sums and the underlined coefficients (divided by 16) are their associated probabilities. Note that the 10 distinct terms in the above multinomial expansion combine to yield just 7 distinct sum terms {2,**3**,4,**5**,6,**7**,8} for a pair of *four-sided dice*.

## 5.2.4.1 Binomial and Trinomial Event Distributions

# Binomial and Trinomial Event Distributions

$n=3$

**Binomial**

$$(a+b)^3 = \sum_{k=0}^{3} \binom{3}{k} a^k b^{3-k}$$

$\#groups = {}^2\mathcal{C}_3 = {}^{2+3-1}C_3 = {}^4C_1 = 4$

$\#permutations \quad {}^2\mathcal{P}_3 = 2^3 = 8$

**n=3 Binomial PMF**

$p_X(x)$

**Tree Expansion/ Grouping**    $a=b=1$    $(2)^3 = \binom{3}{0} + \binom{3}{1} + \binom{3}{2} + \binom{3}{3}$

**Random Variable PMF**    $(p+q)^3 = \binom{3}{0} p^0 q^3 + \binom{3}{1} p^1 q^2 + \binom{3}{2} p^2 q^1 + \binom{3}{3} p^3 q^0$

$a=p$ ; $b=q$ ; $p+q=1$

| | #ways | $2^3=8=$ | 1 | + | 3 | + | 3 | + | 1 |
|---|---|---|---|---|---|---|---|---|---|
| Repeated | #succ | 0 | | 1 | | 2 | | 3 | |
| Indep Trials | #fail | 3 | | 2 | | 1 | | 0 | |

${}^3C_0 \over 8$  ${}^3C_1 \over 8$  ${}^3C_2 \over 8$  ${}^3C_3 \over 8$

**Trinomial**

$n=3$

$$(a+b+c)^3 = \sum_{\substack{all\ r_1, r_2, r_3 \\ r_1+r_2+r_3=3}} \binom{3}{r_1\ r_2\ r_3} a^{r_1} b^{r_2} c^{r_3}$$

$\#groups = {}^3\mathcal{C}_3 = {}^{3+(3-1)}C_3 = (5*4*3)/3! = 10$ terms

$\#permutations = {}^3\mathcal{P}_3 = 3*3*3 = 3^3 = 27$ terms

**Tree Expansion/ Grouping**    $(3)^3 = 27 = \underbrace{\binom{3}{0\,0\,3} + \binom{3}{0\,1\,2} + \cdots + \binom{3}{0\,3\,0} + \cdots + \binom{3}{3\,0\,0}}_{10\ terms\ or\ "groups"}$

$a=b=c=1$

$r_2$   $r_1$

**Random Variable PMF**

$a=p$ ; $b=q$ ; $c=r$:

$p+q+r=1$

$1 = (p+q+r)^3 = \binom{3}{0\,0\,3} p^0 q^0 r^3 + \binom{3}{0\,1\,2} p^0 q^1 r^2 + \cdots + \binom{3}{3\,0\,0} p^3 q^0 r^0$

**Repeated**

**Indep Trials**    *[Bin#1, Bin#2, Bin#3]*    **Occupancy Vectors**

The analysis of repeated independent trials is related to "draws with replacement" that we have previously discussed. Independent trials each have the same probability of success p or failure q=1-p. For many trials it makes sense to talk about the distribution of outcomes which is called a probability distribution. One such distribution is the binomial distribution which is a direct consequence of the binomial expansion and is shown in the figure. The terms in the binomial expansion when normalized to unity yields precisely the probabilities associated with the binomial distribution.

The binomial expansion can be "normalized" by either setting a=b=1 and then dividing by $(1+1)^n$ to yield unity on the LHS and the distribution function on the right or by substituting a=p, b=q=1-p in the binomial expansion to yield $(p+(1-p))^n = 1$ again on the LHS and the distribution on the RHS. The first method holds for p=q=1/2 while the second method holds for all choices of the single trial probability of success p.

Similar remarks hold for the trinomial distribution, although we cannot show a 2-dimensional plot in that case. In the general multinomial case, the expansion of $(p_1+p_2+p_3+...+p_r)^n$, yields coefficient values (multiplicities) given by the multinomial ${}^nC_{i1,i2,\,...ir}$ with index constraint $i_1+i_2+...+i_r = r$. The *number of distinct terms in this expansion* is equal to the number of integer solutions to the index constraint, which is found from *combinations with replacement* to be slash-${}^rC_n = {}^{r+n-1}C_n = {}^{r+n-1}C_n$. (This is also equivalent to the n- balls and r-bins arrangement solution ${}^{r+n-1}C_{r-1}$).

## 5.2.4.2 Trinomial Example

# Trinomial Example

***Randomize n=10 Class Sessions*** into r =3  outcomes

Every Day I spin a wheel which has three

sectors marked with an L , an E and a D.

The sectors are in ***proportion of  7:2:1*** as shown

***What are the "single trial probabilities?"***

| | |
|---|---|
| L = Lecture | $P(L) = p = .7$ |
| E = Exam | $P(E) = q = .2$ |
| D = Dismiss Class | $P(D) = r = .1$ |

***Find Probability*** that in *10 trials* there will be ***exactly*** 8 "L", 2 "E", 0"D"

$$P(8"L", 2"E", 0"D") = \binom{10}{8,2,0}(.7)^8(.2)^2(.1)^0$$

$$= \frac{10!}{8!2!0!}(.7)^8(.2)^2 = .1038$$

***Trinomial expansion***    $^3P_{10} = 3^{10} = 59{,}049$  **Permutations**

$(L + E + D)^{10}$    $^3C_{10} = {}^{12}C_{10} = 66$ **Combinations**

$$(p+q+r)^{10} = \sum_{\substack{j,k,l=0 \\ j+k+l=10}}^{10} \binom{10}{j\,k\,l}p^j q^k r^l = \binom{10}{10\,0\,0}p^{10}q^0r^0 + \binom{10}{9\,1\,0}p^9 q^1 r^0 + \cdots + \binom{10}{0\,0\,10}p^0 q^0 r^{10}$$

$^3C_{10} = {}^{12}C_{10} = 66$ **Terms**

A professor decides what to do in any given class by spinning the wheel having 3 unequal sectors designating *lecture* (L), *exam* (E), and *dismiss class* (D). Each spin is an independent trial with probabilities P(L)=.7, P(E)=.2 and P(D)= .1 as illustrated. Thus the probability distribution generated by the n=10 class sessions is a trinomial and for example the probability that there will be *exactly 8 lectures, 2 exams, and 0 dismissed classes* is easily written down as the term in the distribution having those parameters, *viz.*,

$$P(8L, 2E, 0D) = {}^{10}C_{8,2,0} (.7)^8 (.2)^2(.1)^0 = .1038,$$

which is surprisingly small considering lecture probability P(L)=.7 is raised to the 8th power. However, note that the multinomial coefficients vary widely, *e.g.*, $^{10}C_{8,2,0} = 45$ *versus* $^{10}C_{4,3,3} = 4200$; thus, even with this last large coefficient, the probability

$$P(4L, 3E, 3D) = {}^{10}C_{4,3,3} (.7)^4 (.2)^3(.1)^3 = .00807$$

is even smaller !

The general expression for the trinomial expansion is given in the bottom boxed equation.  There are 59,049 possible tree outputs resulting from 10 draws from 3 objects with replacement and this yields 66 distinct combinations or terms in the trinomial expansion.  We shall see that the sum of terms in any distribution must be unity; since p+q+r =1, it is clear from the trinomial expansion of $(p+q+r)^{10}$ that the resulting terms are the distinct elements of a "trinomial probability distribution".

## 5.2.4.3 Class Sessions Trinomial Table and Plot

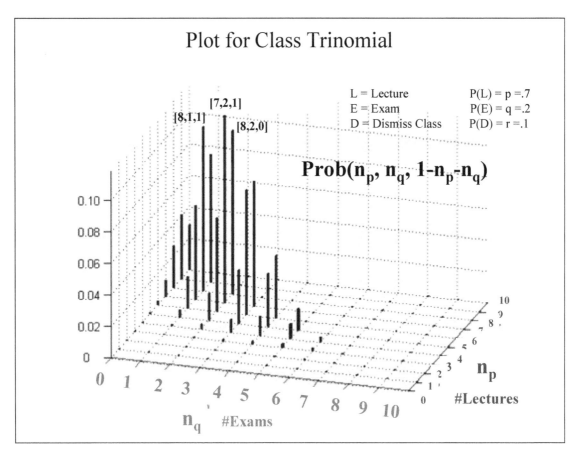

Plot for Class Trinomial

L = Lecture    P(L) = p =.7
E = Exam    P(E) = q =.2
D = Dismiss Class    P(D) = r =.1

$Prob(n_p, n_q, 1-n_p-n_q)$

| $n_q / n_p$ | 0 | 1 | 2 | 3 | 4 | 5 | 6 | 7 | 8 | 9 | 10 |
|---|---|---|---|---|---|---|---|---|---|---|---|
| 0 | 1.00E-10 | 7.00E-09 | 2.21E-07 | 4.12E-06 | 5.04E-05 | 4.24E-04 | 2.47E-03 | 9.88E-03 | 2.59E-02 | 4.04E-02 | 2.82E-02 |
| 1 | 2.00E-09 | 1.26E-07 | 3.53E-06 | 5.76E-05 | 6.05E-04 | 4.24E-03 | 1.98E-02 | 5.93E-02 | 1.04E-01 | 8.07E-02 | 0 |
| 2 | 1.80E-08 | 1.01E-06 | 2.47E-05 | 3.46E-04 | 3.03E-03 | 1.69E-02 | 5.93E-02 | 1.19E-01 | 1.04E-01 | 0 | 0 |
| 3 | 9.60E-08 | 4.70E-06 | 9.88E-05 | 1.15E-03 | 8.07E-03 | 3.39E-02 | 7.91E-02 | 7.91E-02 | 0 | 0 | 0 |
| 4 | 3.36E-07 | 1.41E-05 | 2.47E-04 | 2.31E-03 | 1.21E-02 | 3.39E-02 | 3.95E-02 | 0 | 0 | 0 | 0 |
| 5 | 8.06E-07 | 2.82E-05 | 3.95E-04 | 2.77E-03 | 9.68E-03 | 1.36E-02 | 0 | 0 | 0 | 0 | 0 |
| 6 | 1.34E-06 | 3.76E-05 | 3.95E-04 | 1.84E-03 | 3.23E-03 | 0 | 0 | 0 | 0 | 0 | 0 |
| 7 | 1.54E-06 | 3.23E-05 | 2.26E-04 | 5.27E-04 | 0 | 0 | 0 | 0 | 0 | 0 | 0 |
| 8 | 1.15E-06 | 1.61E-05 | 5.64E-05 | 0 | 0 | 0 | 0 | 0 | 0 | 0 | 0 |
| 9 | 5.12E-07 | 3.58E-06 | 0 | 0 | 0 | 0 | 0 | 0 | 0 | 0 | 0 |
| 10 | 1.02E-07 | 0 | 0 | 0 | 0 | 0 | 0 | 0 | 0 | 0 | 0 |

Explicit plot and table for the previous slide shows detailed structure of the distribution as a function of the number of Lectures $n_p$ and the number of Exams $n_q$. The number of Dismissals is implicit since $n_r = 10 - n_p - n_q$ . It is perhaps surprising that the probability P[L,E, D]: P[8,2,0]= P[8,1,1] =0.104 and that max is P[7,2,1] =0.119.

# 6 Random Variables and Distributions

## 6.1 *Random Variables and Distributions*

# Random Variables and Distributions

1. Experimental Outcomes: RV Distributions
2. Joint, Marginal, & Conditional PMFs & CDFs
3. Mean,Variance, Covariance, & Transformations
4. Common PMFs: Properties & Examples
5. Sums, Convolution, & Moment Generating Function

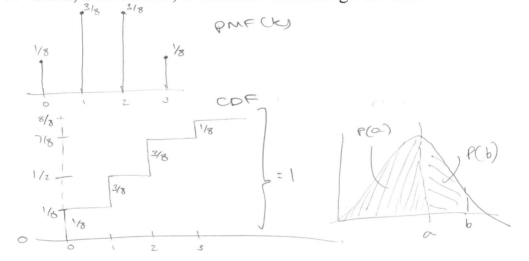

Up to this point we have computed probabilities for single events, but often we are interested in a whole set of events generated in a given sample space. A random variable (RV) is a variable that takes on all possible values describing the outcome of a random experiment. Hence its properties can be used to describe a whole class of random experiments in terms of a single list of probability assignments, a so called *probability mass function* (PMF) or more loosely a *probability density*. The progressive sum as we go through the list of PMF values yields a cumulative distribution function (CDF) or *probability distribution*. Even though the random variable X only takes on discrete values, the CDF is a *quasi-continuous* function that *always starts at zero, never decreases, and always ends at unity*. The CDF plays a central role in probability calculations because it gives probabilities over ranges of values that define compound events of interest.

When more than one RV is needed to describe a set of experiments, it is natural to characterize the outcomes using all the relevant RVs and extend the PMF concept to include joint, conditional, and marginal PMFs. The shape of a distribution says everything there is to know about a RV and the 1st, 2nd, and higher order moments of the distribution describe its shape in ever-increasing detail. A simpler, *albeit* less precise, characterization of a distribution is given in terms of its first two moments which relate directly to its mean and variance. For a pair of random variables we can again characterize the distribution by its first two moments and this leads to two means, two variances, and a covariance expressing their degree of dependence.

This characterization of groups of random experiments leads to a set of common PMFs whose mathematical properties are known and whose application to a broad range of physical problems are well documented. Linear and bi-linear transformations on one or two random variables often lead to PMFs (defined on these new RVs) that are more relevant to a given problem. All these concepts lead to a very rich formulation of probability that helps to crystallize our understanding of physical processes and their associated random variables.

## 6.1.1 Discrete Random Variables (RV) –Key Concepts

# Discrete Random Variables (RV) –Key Concepts

- **Discrete RVs:** A series of measurements of random events
- **Prob Mass Fcn:** (PMF), Joint, Marginal, Conditional PMFs
- **Characteristics:** "Moments:" Mean and Std Deviation
- **Cumulative Distr Fcn:** (CDF) i) Btwn 0 and 1, ii) Non-decreasing
- **Independence** of two RVs
- **Transformations** - Derived RVs
- **Expected Values** (for given PMF)
- **Relationships Btwn two RVs:** Correlations
- **Common PMFs Table**
- **Applications** of Common PMFs
- **Sums & Convolution:** Polynomial Multiplication
- **Generating Function:** Concept & Examples

This slide gives a glossary of some of the key concepts involving random variables (RVs) that will be discussed in detail within this section. Physical phenomena are always subject to some random components, so RVs must appear in any realistic model; an understanding of their statistical properties provides a framework for analysis of many different experiments using the same model. These probability concepts provide a rich set of tools that allow analysis of complex random systems involving several RVs. They also allow us to characterize the transformed distributions resulting from the mathematical equations used to model the system.

At any instant, a RV takes on a single value and represents one sample from the underlying RV distribution defined by its probability mass function (PMF). Often we need to know the probability for some range of values of a RV and this is found by summing the individual probability values of the PMF; thus a cumulative distribution function (CDF) is defined to handle such sums. The CDF formally characterizes the discrete RV in terms of a quasi-continuous function that ranges between [0,1] and which has a unique inverse.

Distributions can also be characterized by single numbers rather than PMFs or CDFs and this leads to concepts of mean values, standard deviations, correlations between pairs of RVs and expected values. There are a number of fundamental PMFs used to describe physical phenomena and these common PMFs will be compared and illustrated through examples. Finally, the relationship between the sum of two RVs and the concept of convolution and the generating function for RVs will be discussed.

## 6.1.2 Random Events Described by Random Variables

# Random Events Described by Random Variables

- **Random Variable (RV)** – numerical assignments to outcomes of "random experiments"
  - *Capital letters* denote RV: "X"

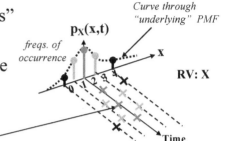

- **Realization of RV** – Single Outcome

  "Experiment: 4 coin Flips: X = #Heads"
  - *Lower case* letters give a specific value of a single outcome: "X = x = 2"

- **Statistics of RV** – Characterize Outcomes $2^4 = 16 = {}^4C_0 + {}^4C_1 + {}^4C_2 + {}^4C_3 + {}^4C_4$

  for "*many random experiments*"
  - for RV "X" : {mean , std deviation}=$\{\mu_X, \sigma_X\}$

$$PMF_X(x) = p_X(x) = P(X = x) \quad ; \quad 0 \le p_X(x) \le 1$$

- **Probability Mass Function (PMF)** – Characterizes Outcomes for "many random experiments"

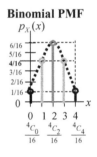

The top figure illustrates the fundamental idea that (at any instant in time) a RV is a single realization (sample) from the underlying RV distribution defined by its probability mass function (PMF). The standard practice is to denote the RV by a capital letter such as X and its realization (instantaneous sample) by the corresponding lower case letter x, so that the statement "X=x" means the RV X takes on the specific value x(= 3.5 say). Note that "X=x" *is not an equality*; rather it says the RV takes on the value "x" in this particular experiment.

The PMF for a binomial RV X is specified by the binomial distribution b(x; n, p) for x draws from n with single trial probability of success p. This PMF is a discrete function of the RV X representing the number of successes "x"; it is written in various notations as follows: $PMF_X(x)$ =Prob[X = x]= $p_X(x)$= b(x;n,p). The figure on the lower right shows a plot of this discrete function in the case of "x" draws from n = 4 with p = .5, written explicitly as

$$p_X(x) = b(x; 4, 0.5) = {}^4C_x (.5)^x (1-.5)^{4-x} = {}^4C_x (.5)^4 \text{ for x=0,1,2,3,4.}$$

The dotted curve shows the "shape of a continuous curve" through the discrete $PMF_X(x)$ values which previews the linkage between the Binomial distribution and its *large n* continuous limit, the Gaussian distribution.

## 6.2  2-Dimensional "Joint" PMF($d_1,d_2$):  4-Sided Pair of Dice

# 2-Dimensional "Joint" PMF($d_1,d_2$) : 4-Sided Pair of Dice

Fair 4-sided dice thrown twice: Every outcome ($d_1,d_2$) equally likely.

The Probability Mass Function **PMF is constant = 1/16**

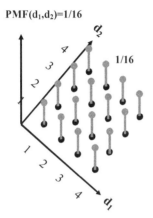

PMF($d_1,d_2$)=1/16

1/16

**$D_1$ & $D_2$  are RVs   Exptl Outcomes for each Die**

**Uniform PMF($d_1,d_2$)**

$$\sum_{d_1,d_2} p_{D_1,D_2}(d_1,d_2)=1 \quad ; \quad 0 \le p_{D_1,D_2}(d_1,d_2) \le 1$$

The sample space for a pair of 4-sided dice is described by points ($d_1$, $d_2$) in a plane corresponding to the 16 possible realizations of the pair of discrete random variables $D_1$, $D_2$. A PMF defined as a function of two (or more) independent RVs is called a "joint PMF" and is written as $PMF_{D_1,D_2}(d_1d_2)$ or alternately $p_{D1,D2}(d_1d_2)$. Since each die takes on face values $d_1,d_2=\{1,2,3,4\}$ independently, the resulting joint PMF factors into the product of two one dimensional PMFs as $p_{D1,D2}(d_1d_2) = p_{D1}(d_1) * p_{D2}(d_2)$  The PMF  $p_{D1,D2}(d_1d_2)$ can be plotted in the $3^{rd}$ dimension perpendicular to the $d_1$-$d_2$ plane as 16 sticks of equal height in the case of fair dice with equally likely outcomes. Thus the PMF $p_{D1,D2}(d_1d_2)$ is a uniform discrete function of coordinates ($d_1$, $d_2$). (This would be represented by a plane at z=1/16 unit in the continuous case.)

## 6.2.1 Transformation of Sample Space: Sum and Difference

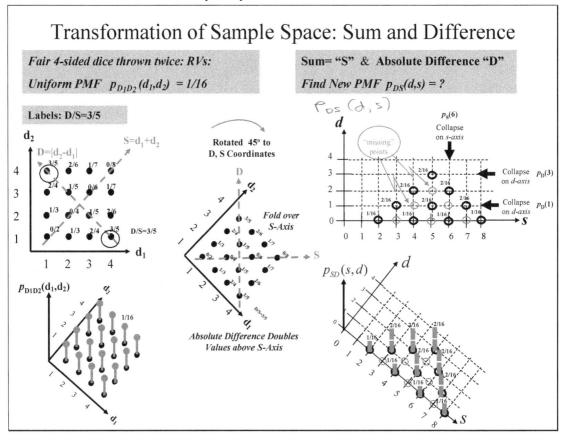

In the game with 4-sided dice, we are interested in the distribution of the sum random variable $S = D_1 + D_2$, $p_S(s)$ and not the joint distribution $p_{D1,D2}(d_1 d_2)$. This slide and several to follow illustrate the procedure for obtaining the desired "marginal" (or collapsed) distribution $p_S(s)$. In the process, we shall develop the relationship between distributions under transformation of coordinates, and define conditional, and marginal distributions involving the pair of RVs $\{D_1, D_2\}$.

We start with the 2- and 3-dimensional PMF representations for equally likely outcomes of 1/16 as shown on the left. Recall that the points $(d_1, d_2)$ for dice outcomes may alternately be expressed by points $(s, d)$ their sum and difference coordinates, where $s = d_1 + d_2$ and $d = d_2 - d_1$. These coordinate axes are shown in the top left figure where the sum and difference take on values: $s=\{2,3,4,5,6,7,8\}$ and $d=\{-3,-2,-1,0,1,2,3\}$. We consider a slightly different transformation $s = d_1 + d_2$ and $|d| = |d_2 - d_1|$ and now the absolute difference $|d|$ takes on only 4 values $\{0,1,2,3\}$; this has the effect of doubling the probability values of $\{1,2,3\}$ by folding over the negative difference values thereby doubling them. If we label each point in this figure by the "$|d|/s$" values we see for example that the points $(d_1, d_2)=(1, 4)$ and $(d_1, d_2)=(4, 1)$ at opposite corners of the grid are both now labeled with $|d|/s = 3/5$. Labeling all points in this manner and rotating the figure clockwise 45° so D is up and S is to the right (central figure) we have found the new joint distribution $p_{SD}(s, |d|)$ as illustrated in the two right figures where points are now labeled by their $(s, |d|)$ values. Note that the new distribution has doubled the positive d values to 2/16 each and that certain coordinate points, *e.g.*, $(s, |d|)=(3, 0)$ are not occupied (green). The marginal distribution $p_S(s)$, defined as the sum of the joint distribution $p_{SD}(s, |d|)$ over all $|d|$ values, is easily picked off the upper right figure by collapsing values down along the s-axis. Similarly, the distribution $p_D(|d|)$ defined as the sum of the joint distribution $p_{SD}(s,|d|)$ over all s-values. The table and graphic on the next slide gives a clear illustration of the collapse (sum) in one dimension which produces the two marginal distributions.

## 6.2.2 Table of Joint, Marginal, and Conditional PMFs 4-Sided Dice

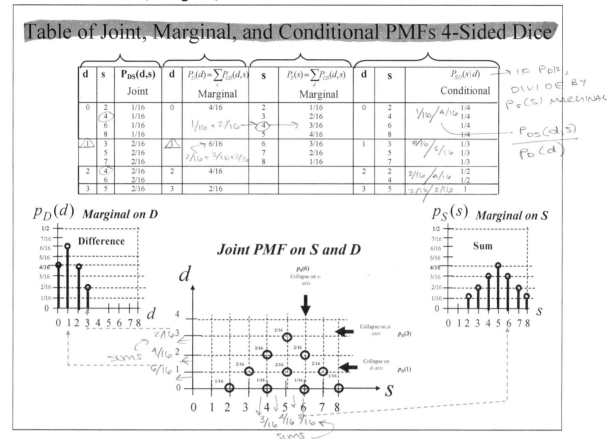

The joint distribution $p_{DS}(d,s)$ from the last slide is plotted here together with the two marginal distributions $p_D(d)$, and $p_S(s)$ whose values are given in the table of the last slide. Starting with the center figure for the joint distribution $p_{DS}(d,s)$, we see that "projecting and summing" the joint probabilities horizontally left onto the d-axis yields the magnitudes of marginals $p_D(d)$ in the left figure; similarly "projecting and summing" the joint probabilities vertically down onto the s-axis yields the magnitudes of marginals $p_S(s)$ in the right figure.

The large table actually represents a summary of 4 separate probability tables, namely, (i) joint probability $p_{DS}(d,s)$, (ii) marginal $p_D(d)$, (iii) marginal $p_S(s)$, and (iv) conditional $p_{S\mid D}(s \mid d)$. The first three tables are taken directly from the three bottom figures. We note that a more natural display of the joint probability would be a 4x7 matrix, but in order to display the tables together we have only written down the non-zero terms of the joint distribution and have done that in the following linear manner. We first display row "d=0" and columns s={2,4,6,8}, then row "d=1" columns s={3,5,7}, row "d=2" columns s={4,6}, and finally row "d=3" columns s={5}. All other values are zero.

The last "sub-table" for the conditional probability $p_{S\mid D}(s \mid d)$ is constructed directly from the joint distribution by using the definition $p_{S\mid D}(s \mid d) = p_{SD}(s , d) / p_D(d)$. Thus for d=0 we find

$p_{S\mid D}(s=\{2,4,6,8\} \mid d=0) = p_{S\,D}(s=\{2,4,6,8\} , d=0) / p_D(d=0)$
$= \{1/16,1/16,1/16,1/16\} / \{4/16\} = \{1/4,1/4,1/4,1/4\}$

which gives the first four entries of the conditional table; similarly

$p_{S\mid D}(s=\{3,5,7\} \mid d=1) = p_{SD}(s=\{3,5,7\},d=1) / p_D(d=1) = \{2/16,2/16,2/16\} /\{6/16\}= \{1/3,1/3,1/3\}$ which gives the next three entries.

## 6.2.3 Compute Conditional PMF $P_{S|D}(s|d)$ and Reconstruct $P_{SD}(s,d)$

# Compute Conditional PMF $P_{S|D}(s|d)$ and Reconstruct $P_{SD}(s,d)$

| d | s | $P_{DS}(d,s)$ | d | $P_D(d)=\sum_s P_{DS}(d,s)$ | s | $P_S(s)=\sum_d P_{DS}(d,s)$ | d | s | $P_{S|D}(s|d)$ |
|---|---|---|---|---|---|---|---|---|---|
| | | Joint | | Marginal | | Marginal | | | Conditional |
| 0 | 2 | 1/16 | 0 | 4/16 | 2 | 1/16 | 0 | 2 | 1/4 |
| | 4 | 1/16 | | | 3 | 2/16 | | 4 | 1/4 |
| | 6 | 1/16 | | | 4 | 3/16 | | 6 | 1/4 |
| | 8 | 1/16 | | | 5 | 4/16 | | 8 | 1/4 |
| 1 | 3 | 2/16 | 1 | 6/16 | 6 | 3/16 | 1 | 3 | 1/3 |
| | 5 | 2/16 | | | 7 | 2/16 | | 5 | 1/3 |
| | 7 | 2/16 | | | 8 | 1/16 | | 7 | 1/3 |
| 2 | 4 | 2/16 | 2 | 4/16 | | | 2 | 4 | 1/2 |
| | 6 | 2/16 | | | | | | 6 | 1/2 |
| 3 | 5 | 2/16 | 3 | 2/16 | | | 3 | 5 | 1 |

**Conditional Probability $P_{S|D}(s|d) = P_{DS}(d,s) / P_D(d)$**

$p_{S|D}(s=\{2,4,6,8\}|\,d=0) = p_{SD}(s=\{2,4,6,8\},d=0) / p_D(d=0)$

$= \{1/16,1/16,1/16,1/16\} / \{4/16\} = \{1/4,1/4,1/4,1/4\}$

$p_{S|D}(s=\{3,5,7\}|\,d=1) = p_{SD}(s=\{3,5,7\}|\,d=1) / p_D(d=1)$

$= \{2/16,2/16,2/16\} / \{6/16\} = \{1/3,1/3,1/3\}$

**Reconstruct Joint Probability $P_{DS}(d,s)= P_{S|D}(s|d)\, P_D(d)$**

$p_{DS}(d=0, s=\{2,4,6,8\}) = p_{S|D}(s=\{2,4,6,8\}|\,d=0)*p_D(d=0)$

$= \{1/4,1/4,1/4,1/4\} *\{4/16\} = \{1/16,1/16,1/16,1/16\}$

$p_{DS}(d=1, s=\{3,5,7\}) = p_{S|D}(s=\{3,5,7\} |\, d=1) *p_D(d=1)$
$= \{1/3,1/3,1/3\}* \{6/16\} = \{2/16,2/16,2/16\}$

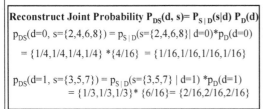

**From Joint Probability obtain both marginals by summing over S or D**

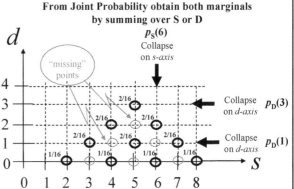

It should be noted that when we obtain marginals by collapsing onto the d-axis or the s-axis we lose information about the structure of the joint distribution and there is no way to "recover" it from the marginals. That is, we cannot reconstruct the joint distribution from the pair of marginals.

In order to construct the joint distribution $p_{S D}(s, d)$, we must have one conditional $p_{S|D}(s \mid d)$ and one marginal $p_D(d)$; for then we construct the joint distribution as $p_{S D}(s, d) = p_{S|D}(s \mid d)* p_D(d)$.

Thus, for d=0 we find

$p_{SD}(d=0, s=\{2,4,6,8\}) = p_{SD}(s=\{2,4,6,8\}|d=0)*p_D(d=0) = \{1/4,1/4,1/4,1/4\}*\{4/16\}$
$=\{1/16,1/16,1/16,1/16\}$

which gives the first four entries for d=0 in the joint probability table; similarly

$p_{s,D}(d=1, s=\{3,5,7\}) = p_{SD}(s=\{3,5,7\}|d=1) \;/\; p_D(d=1) = \{1/3,1/3,1/3\}*\{6/16\}= \{2/16,2/16,2/16\}$

which gives the next three entries.

## 6.3 *Min-Max Derived Random Variables*

# Min-Max Derived Random Variables

- Derived Random Variables: All have different PMFs – go to Original Sample Space
  - $L=\min(D_1,D_2)$    $U=\max(D_1,D_2)$
- **$L=\min(D_1,D_2)$**

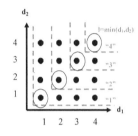

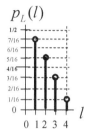

| L=min(D₁,D₂) | |
|---|---|
| **L** | **Pr(L=l)** |
| 1 | 7/16 |
| 2 | 5/16 |
| 3 | 316 |
| 4 | 1/16 |

- **$U=\max(D_1,D_2)$**

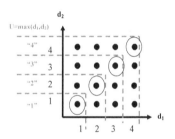

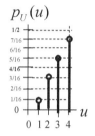

| U=max(D₁,D₂) | |
|---|---|
| **U** | **Pr(U=u)** |
| 1 | 1/16 |
| 2 | 3/16 |
| 3 | 5/16 |
| 4 | 7/16 |

For a pair 4-sided dice we make the coordinate transformation from $\{D_1, D_2\}$ to $\{L, U\}$, where $L=\min(D_1, D_2)$ and $U=\max(D_1, D_2)$ to define the lower L and upper U order statistics. The effect of the coordinate transformation is most easily understood by looking at the "corner shapes" shown in the $\{D_1, D_2\}$ dice representation.

In the upper figure, we display surfaces of constant L-value as the outward facing corner surfaces shown. The minimum value "L=1" starts on the diagonal point $(1,1)$ contains all points for which one coordinate is fixed at 1, and the other ranges over $\{1,2,3,4\}$ and hence contains the 7 points $\{(1,1), (1,2),(1,3),(1,4),(2,1),(3,1),(4,1)\}$; accordingly the value 7/16 appears in the table and in the PMF plot $p_L(\ell)$ for L=1. Similarly, the surface "L=2" starts at the diagonal point $(2,2)$ and contains all points for which either coordinate is $\geq 1$ and hence contains the 5 points $\{(2,2), (2,3),(2,4),(3,2),(4,2)\}$; accordingly the value 5/16 appears in the table and in the PMF plot $p_L(\ell)$ for L=2.

In the lower figure, we display surfaces of constant U-value as the inward facing corner surfaces shown. The maximum value "U=4" starts on the diagonal point $(4,4)$ contains all points for which either coordinate is $\leq 4$ and hence contains the 7 points $\{(4,4), (4,3),(4,2),(4,1),(3,4),(2,4),(1,4)\}$; accordingly the value 7/16 appears in the table and in the PMF plot $p_U(u)$ for U=4. Similarly, the surface "U=3" starts at the diagonal point $(3,3)$ and contains all points for which either coordinate is $\leq 3$ and hence contains the 5 points $\{(3,3), (3,2),(3,1),(2,3),(1,3)\}$; accordingly the value 5/16 appears in the table and in the PMF plot $p_U(u)$ for U=3.

Comparing the PMFs for L and U we see that they contain the same values but have opposite behavior; specifically $p_L(\ell)$ decreases with "$\ell$" whereas $p_U(u)$ increases with "u" . Thus by inspection, we immediately have the two marginal probabilities, but recall that we cannot construct the joint PMF $p_{LU}(\ell,u)$ from the two marginals.

## 6.3.1 Construction of Joint PMF p$_{UL}$(u,l) and its Sample Space

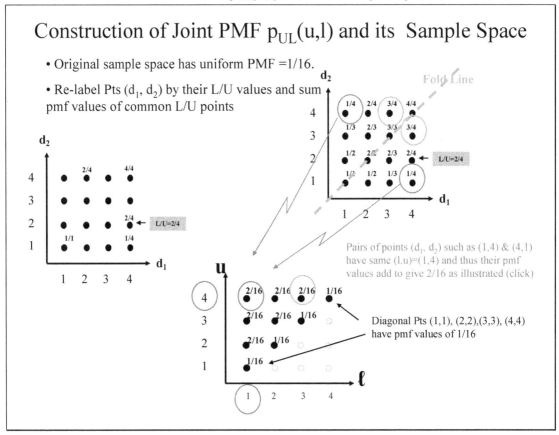

In order to construct the joint PMF $p_{LU}(\ell,u)$, we follow the procedure we used to find the joint PMF $p_{SD}(s,d)$ for the sum difference coordinates (S,D) (except there is no rotation between coordinates in this case.) Accordingly, we re-label the points in the original {$D_1$, $D_2$} sample space, this time with their L/U-values. For clarity, this is illustrated for a few points in the left figure and then for all points in the right figure. The right figure also shows the diagonal fold line about which we have identical values as indicated by the two green circles and the two red values.

In the right figure we see that the two distinct red points $(d_1,d_2)=(1,4)$ and $(d_1,d_2)=(4,1)$ correspond to a single red point in {L, U}coordinates $(\ell, u)=(1,4)$ and they sum to a value of 2/16 in the middle figure; similarly $(d_1,d_2)=(3,4)$ and $(d_1,d_2)=(4,3)$ correspond to a single green point in{L, U}coordinates $(\ell, u) = (3,4)$ and sum to a value of 2/16. Thus the PMF values in the{L, U}coordinates are doubled to 2/16 above the L=U diagonal, remain at 1/16 on the diagonal, and are zero below the diagonal as shown in the middle figure displaying the desired $p_{LU}(\ell, u)$.

On the next slide we show a composite table for joint, marginals, and conditional probabilities, as well as plots of the joint distribution $p_{LU}(\ell, u)$ and its associated marginals.

## 6.3.2 Min-Max Joint, Marginal, and Conditional PMFs

# Min-Max Joint, Marginal, and Conditional PMFs

| l | u | $P_{UL}(u,l)$ Joint | | l | $P_L(l)=\sum P_{UL}(u,l)$ Marginal | | u | $P_U(u)=\sum P_{UL}(u,l)$ Marginal | | given l | u | $P_{UL}(u\mid l)$ Conditional |
|---|---|---|---|---|---|---|---|---|---|---|---|---|
| 1 | 1 | 1/16 | | 1 | 7/16 | | 1 | 1/16 | | 1 | 1 | (1/16)/(7/16) = 1/7 |
|   | 2 | 2/16 | | | | | | | | | 2 | (2/16)/(7/16) = 2/7 |
|   | 3 | 2/16 | | | | | | | | | 3 | (2/16)/(7/16) = 2/7 |
|   | 4 | 2/16 | | | | | | | | | 4 | (2/16)/(7/16) = 2/7 |
| 2 | 2 | 1/16 | | 2 | 5/16 | | 2 | 3/16 | | 2 | 2 | (1/16)/(5/16) = 1/5 |
|   | 3 | 2/16 | | | | | | | | | 3 | (2/16)/(5/16) = 2/5 |
|   | 4 | 2/16 | | | | | | | | | 4 | (2/16)/(5/16) = 2/5 |
| 3 | 3 | 1/16 | | 3 | 3/16 | | 3 | 5/16 | | 3 | 3 | (1/16)/(3/16) = 1/3 |
|   | 4 | 2/16 | | | | | | | | | 4 | (2/16)/(3/16) = 2/3 |
| 4 | 4 | 1/16 | | 4 | 1/16 | | 4 | 7/16 | | 4 | 4 | (1/16)/(1/16) = 1 |

Conditional Probability $P_{UL}(u\mid l) = P_{UL}(u, l) / P_L(l)$

$p_{U\mid L}(u=\{1,2,3,4\} \mid l=1) = p_{UL}(u=\{1,2,3,4\}, l=1) / p_L(l=1)$
$= \{1/16,2/16,2/16,2/16\} / \{7/16\}=$
$\{1/7,2/7,2/7,2/7\}$
$p_{U\mid L}(u=\{2,3,4\} \mid l=2) = p_{UL}(u=\{2,3,4\}, l=2) / p_L(l=2)$
$= \{1/16,2/16,2/16\} / \{5/16\}= \{1/5,2/5,2/5\}$

The joint distribution $p_{UL}(u, l)$ from the last slide is plotted here together with the two marginal distributions $p_L(l)$, and $p_U(u)$ previously shown. Starting with the center figure for the joint distribution $p_{UL}(u, l)$ we see that "projecting and summing" the joint probabilities horizontally left onto the u-axis yields the magnitudes of marginals $p_U(u)$ in the left figure; similarly "projecting and summing" the joint probabilities vertically down onto the l-axis yields the magnitudes of marginals $p_L(l)$, in the upper figure. A 3-dimensional representation of $p_{UL}(u, l)$ is also shown on the right.

The large table actually represents a summary of 4 separate probability tables, namely, (i) joint probability $p_{UL}(u, l)$, (ii) marginal $p_L(l)$, (iii) marginal $p_U(u)$, and (iv) conditional $p_{UL}(u\mid l)$. The first three tables are taken directly from the three bottom figures. We note that a more natural display of the joint probability would be a 4 x 4 matrix, but in order to display the tables together we have only written down the non-zero terms of the joint distribution and have done that in the following linear manner. We first display row "l =1" and columns u={1,2,3,4}, then row "l =2" columns u={2,3,4}, row "l =3" columns u={3,4}, and finally row "l =4" column u={4}. All other values are zero.

The last "sub-table" for the conditional probability $p_{UL}(u\mid l)$ is constructed directly from the joint distribution by using the definition $p_{U\mid L}(u\mid l) = p_{U,L}(u, l)/ p_L(l)$. Thus for l =1 we find

$p_{U\mid L}(u=\{1,2,3,4\}\mid l =1) = p_{UL}(u=\{1,2,3,4\}, l =1) / p_L(l =1)$
$= \{1/16,2/16,2/16,2/16\} / \{7/16\}= \{1/7,2/7,2/7,2/7\}$

which gives the first four entries of the conditional table; similarly

$p_{U\mid L}(u=\{2,3,4\}\mid l =2) = p_{UL}(u=\{2,3,4\}, l =2) / p_L(l =2) = \{1/16,2/16,2/16\} / \{5/16\}= \{1/5,2/5,2/5\}$
which gives the next three entries.

## 6.4 Independence of Two Random Variables/ Derived RVs

Independence of Two Random Variables/ Derived RVs

- X and Y Indep RVs iff $\quad p_{X,Y}(x,y) = p_X(x) \cdot p_Y(y)$ for all $x,y$
- Conditional Indep. $\quad p_{X,Y|A}(x,y|a) = p_{X|A}(x|a) \cdot p_{Y|A}(y|a)$ for all $x,y$
- If Indep Condition "Fails" for one pair of values

  Then not independent $\qquad p_{S,D}(2,0) \overset{?}{=} p_S(2) \cdot p_D(0)$

  Condition fails so S & D are **NOT** Indep RVs

  Built in dependence btwn S & D since $\qquad p_{S,D}(2,0) = \dfrac{1}{16}$ ; $p_S(2) = \dfrac{1}{16}$ ; $p_D(0) = \dfrac{4}{16}$

  for example: only sums of equal numbers

  can have zero differences $\qquad \dfrac{1}{16} \neq \dfrac{1}{16} \times \dfrac{4}{16}$ **Fails**

- Derived Random Variables:
  - $S = D_1 + D_2 \qquad\qquad D = |D_2 - D_1|$
  - $U = D_1^2 \qquad\qquad\quad V = D_1 \cosh(D_2),$
  - $W = D_1^2 + D_2 \qquad\quad L = \min(D_1, D_2)$

The definitions of independence and conditional independence for events are now replaced by equivalent ones for random variables. Inasmuch as the PMF embodies the properties of a RV for all realizations (samples), the concepts of independence and conditional independence for random variables are expressed as properties of their distributions. Thus we define independence and conditional independence for joint distributions on the random variables X and Y as

$$p_{XY}(x,y) = p_X(x) * p_Y(y)$$
$$p_{XY|A}(x,y|A) = p_{X|A}(x|A) * p_{Y|A}(y|A)$$

where it is important to understand that for these relations to hold for RVs, *they must hold for all realizations* X=x, Y=y. As an example, we can test whether S and D, the transformed sample space for 4-sided dice are independent by testing *all samples; it only takes one failure to make them dependent.* Thus, testing $p_{SD}(s=2, d=0) = 1/16$ against the product $p_S(s=2) * p_D(d=0) = 1/16 * 4/16$ we show by a *single exception* to the test, that S and D are *not independent.* The reason for this is that there is an obvious *dependence* between sums and differences because only sums of equal numbers can have a difference d=0. Comparing the pair $(d_1,d_2)=(2,2)$ which has s=4 and d=0 with the pair $(d_1,d_2)=(3,1)$ which has s=4 and d=1 clearly shows $(d_1,d_2)$ cannot range freely if we fix d=0.

Physical engineering models can transform the input RV distributions through the mathematical models as shown by several transformations on the slide. The transformations of $\{D_1, D_2\}$ to $\{S, D\}$, or to $\{L, U\}$ have been discussed at length; the $\{min, max\}$ transformation from $\{D_1, D_2\}$ to $\{L, U\}$ is the simplest of a class of RVs known as "order statistics," which can be extended to any number of RVs, e.g. { minimum, middle1, maximum}, { minimum, middle1 middle2, maximum}, *etc.* .

## 6.4.1 Joint PMF Conditioned on a Composite Event A

# Joint PMF Conditioned on a Composite Event A

$$p_{X,Y|A}(x,y \mid A) = \begin{cases} \dfrac{p_{X,Y}(x,y)}{P(A)} & , \quad x,y \in A \\ 0 & , \quad \text{Otherwise} \end{cases}$$

*Event (x,y) is counted only if it is in A; otherwise its probability is set to zero.*

*Renormalize by dividing by P(A)*

*Verify normalization of probability*

$$\sum_x \sum_y P_{X,Y|A}(x,y \mid A) = \sum_{(x,y) \in A} \frac{p_{X,Y}(x,y)}{P(A)} + \sum_{(x,y) \notin A} \frac{0}{P(A)} = \frac{1}{P(A)} \sum_{(x,y) \in A} p_{X,Y}(x,y) = \frac{P(A)}{P(A)} = 1$$

*Example: Pair of 4-sided Dice*

**P(B)= sum all in B**     **P(A)= sum all in A**

$$P(B) = \frac{2}{16} + \frac{1}{16} = \frac{3}{16} \qquad P(A) = \frac{2}{16} + \frac{2}{16} + \frac{1}{16} + \frac{1}{16} = \frac{6}{16}$$

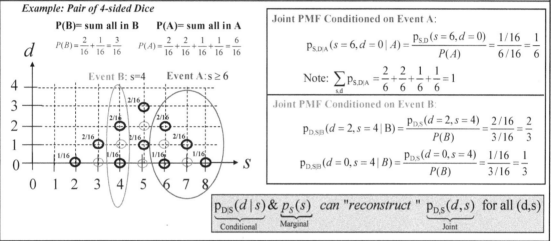

**Joint PMF Conditioned on Event A:**

$$p_{S,D|A}(s=6, d=0 \mid A) = \frac{p_{S,D}(s=6, d=0)}{P(A)} = \frac{1/16}{6/16} = \frac{1}{6}$$

$$\text{Note: } \sum_{s,d} p_{S,D|A} = \frac{2}{6} + \frac{2}{6} + \frac{1}{6} + \frac{1}{6} = 1$$

**Joint PMF Conditioned on Event B:**

$$p_{D,S|B}(d=2, s=4 \mid B) = \frac{p_{D,S}(d=2, s=4)}{P(B)} = \frac{2/16}{3/16} = \frac{2}{3}$$

$$p_{D,S|B}(d=0, s=4 \mid B) = \frac{p_{D,S}(d=0, s=4)}{P(B)} = \frac{1/16}{3/16} = \frac{1}{3}$$

$$\underbrace{p_{D|S}(d \mid s)}_{\text{Conditional}} \& \underbrace{p_S(s)}_{\text{Marginal}} \text{ can "reconstruct " } \underbrace{p_{D,S}(d,s)}_{\text{Joint}} \text{ for all (d,s)}$$

A joint PMF $p_{XY}(x,y)$ may be conditioned on any subset of its domain such as events A or B shown in the figure; thus, conditioning it on the event A gives a new joint PMF

$$p_{XY|A}(x,y|A) = p_{XYA}(x,y,A)/p_A(A)$$

The joint notation (x,y,A) is equivalent to an intersection of the joint event (x, y) with event A; it can be written "without the intersection with A", by instead adding an equivalent word statement. Thus the boxed definition at the top of the page reads "$p_{XY|A}(x,y|A)$ equals $p_{XY}(x,y)/p_A(A)$ for points (x, y) in A and is zero otherwise." Defined in this way, the sum of $p_{XY|A}(x,y|A)$ over all (x,y) (even those not in A) is properly normalized to unity as displayed in the center panel. In the bottom panel, this conditional PMF definition is applied to the two events A and B within the sum-difference sample space for a pair of 4-sided dice. Thus, for event A, the conditional probability $p_{SD|A}(s=6, d=0|A)$ is found by taking the original density value $p_{SD}(s=6, d=0) = 1/16$ and dividing it by $P(A) = 6/16$ to yield a probability of 1/6. Alternately, this result can be "read" directly off the figure by restricting ourselves to the 6 *equally likely points* that comprise event A; the probability for the *single point* (s=6, d=0) is clearly just 1/6. The probabilities for events conditioned on event B follow in a similar manner.

## 6.5  The Cumulative (Probability) Distribution Function (CDF)

# The Cumulative (Probability) Distribution Function (CDF)

- ## Cumulative probability assigned to event $X \leq x$

$$F_X(x) = P[X \leq x] = \sum_{\xi \leq x} p_X(\xi)$$

$$F_{XY}(x,y) = P[X \leq x, Y \leq y] = \sum_{\eta \leq y} \sum_{\xi \leq x} p_{X,Y}(\xi, \eta)$$

DO NOT NEED TO KNOW

Non-decreasing function

Continuous from the right $\lim_{\varepsilon \to 0} F_X(x_i + \varepsilon) = F_X(x_i)$

Jump discontinuity from the left: $F_X(x_i) - F_X(x_{i-1}) = p_X(x_i)$

*Jump Discontinuities are the PMF Values*

$p_X(x=5) = F_X(5) - F_X(4)$
= 10/16-6/16 = 4/16

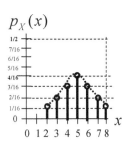

CDF function obtained by taking cumulative sum of $p_X(x)$ for all x less than 1, then less than 2, etc. ...

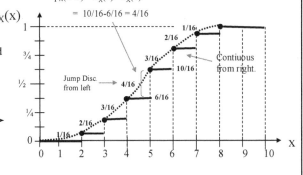

---

The CDF formally characterizes the discrete RV in terms of a quasi-continuous function that takes on values only in the interval [0,1] and it has a unique inverse. The importance of the concept of cumulative probability distribution (CDF) and its fundamental properties cannot be over emphasized.

The CDF function $F_X(x)$ is defined as the cumulative sum of probabilities $p_X(x)$ from X= -∞ to X=x and therefore represents the probability that X ≤x, or explicitly, $F_X(x) = P[X \leq x]$.  The concept is easily generalized to two (or more) random variables X, Y as $F_{XY}(x,y) = P[X \leq x, Y \leq y]$ where the sum is now in 2-dimensions from (-∞,-∞) to the point (x, y); this is in fact the sum of all probability masses within the maximum or inward-facing corner surface at that point (See Slide# 6-9 on the min-max coordinates).

Since the probability $p_X(x)$ is a number in [0,1], the cumulative probability (i) starts at a value of zero, (ii) can never decrease (it can stay constant), and (iii) always ends at a value of 1.  It is also quasi-continuous from the right and may have jump discontinuities from the left as shown in the figure.  Most importantly it is an invertible function; this means that, given a specific value $y_0$=3/16 =$F_X(x)$, we can always find a unique discrete value $x_0$ =3 = $F_X^{-1}(y_0=3/16)$ since changes in cumulative values only occur at the jump discontinuities.

The discrete distribution shown on the left bottom yields the CDF function $F_X(x)$ obtained by summing all contributions for X ≤0 then ≤1, then ≤2 *etc.*, until all probability mass has been accumulated and sums to 1. Thus, specifically $F_X(x) = 0$ from -∞ up to X=2 where it jumps to 1/16; it remains at 1/16 until it reaches X=3 at which point it accumulates a jump of 2/16 to reach a value of 3/16; again it remains at 3/16 until X=4 at which point it accumulates a jump of 3/16 to reach a value of 5/16; it continues in this manner until it attains a value of 1 at X=8 and it remains at 1 thereafter.

## 6.5.1 Expected Value of a RV

# Expected Value of a RV

- **Expected Value for Function g(X) of Single RV: X**

$$E[g(X)] \equiv \sum_x g(x) \cdot p_X(x) \quad \text{Expected Value of g(X)}$$

$$E[g(X) \mid A] \equiv \sum_x g(x) \cdot p_{X|A}(x) \quad \text{Conditional Expected Value of g(X)}$$

- **Expected Value for Sum of Two RVs: X+Y**

$$E[X+Y] \equiv \sum_x \sum_y (x+y) \cdot p_{X,Y}(x,y) = E[X] + E[Y] \quad \text{Independence \textbf{NOT} Required !!}$$

$$E[X+Y] = \sum_x x \cdot \underbrace{\sum_y p_{X,Y}(x,y)}_{\text{Marginal: } p_X(x)} + \sum_y y \cdot \underbrace{\sum_x p_{X,Y}(x,y)}_{\text{Marginal: } p_Y(y)} = E[X] + E[Y]$$

- **Expected Value for Prod of Two RVs: XY**

$$E[X \cdot Y] \equiv \sum_x \sum_y (x \cdot y) \cdot p_{X,Y}(x,y) = E[X] \cdot E[Y] \quad \text{Independence \textbf{IS} Required !!}$$

If X, Y are independent RVs, Then $p_{X,Y}(x,y) = p_X(x) \cdot p_Y(y)$ and we have

$$E[X \cdot Y] = \underbrace{\sum_x x \cdot p_X(x)}_{\equiv E[X]} \cdot \underbrace{\sum_y y \cdot p_Y(y)}_{\equiv E[Y]} = E[X] \cdot E[Y]$$

The expected value E[X] of a random variable X is defined to be the "PMF-weighted average" over all values of X for which its PMF is non-zero; it represents a predicted or theoretical mean that can be compared to the *sample average* of a large set of experimental observations (*realizations* X=x.) For a general function g(X), the expectation value E[g(X)] takes the "PMF-weighted average" of that function *expressed in terms of the random variable* X; *e.g.*, for $g(x)=x^2+5x$, we have $E[X^2+5X]$ expressed as the sum of terms $\sum(x^2+5x)*p_X(x)$ over all values for which $p_X(x)\neq0$.

The expected value of X conditioned on a second random variable A, E[X|A] (reduced sample space) is naturally defined as the "conditional PMF-weighted average" over all values of X in the reduced sample space for which the conditional PMF is non-zero. For example, $E[X^2+5X|X>3]$ is expressed as the sum of terms $\sum(x^2+5x)*p_{X|X>3}(x|x>3)$ over all values for which $p_{X|X>3}(x|x>3) \neq0$.

The expected value E[Z] of the *sum* of two random variables Z=X+Y, is easily shown to separate into the sum E[Z] =E[X+Y]=E[X]+E[Y] *without any restrictions*, such as a requirement that X and Y be independent. On the other hand, the expected value E[Z] of the *product* of two random variables Z=X*Y factors into a product of their individual expectations E[Z] =E[X*Y]= E[X]*E[Y] only when X and Y *are independent*.

In the general case of many random variables, $X_1$, $X_2$, ..., $X_n$, the expectation of their sum always separates into a sum of the individual expectations, while the expectation of their product only separates into a product of individual expectations when the RVs are *mutually independent*.

Random Variables and Distributions

## 6.6 *Numbers That Characterize RVs: Mean, Covariance, Correlation*

Numbers That Characterize RVs: Mean, Covariance, Correlation

- **Single Random Variable: Approximately Characterized by two numbers**

  1st moment (mean)     2nd "centered" moment (variance = (stdev)$^2$ )

  $$\mu_x \equiv E[X]$$

  $$\text{var}(X) \equiv E[(X-\mu_X)\cdot(X-\mu_X)] = E[X^2 - 2\mu_X \cdot X + \mu_x^2]$$
  $$\sigma_X^2 = E[X^2] - E[X]^2 = E[X^2] - \mu_X^2$$

- **Covariance of Two RVs: Cov(X,Y) - Measures "Centered Dependency"**

  $$\text{cov}(X,Y) \equiv E[(X-\mu_X)\cdot(Y-\mu_Y)] \; ; \; \mu_X \equiv E[X] \; ; \; \mu_Y \equiv E[Y]$$

  $$\text{cov}(X,Y) = E[XY - \mu_X Y - \mu_Y X + \mu_X\mu_Y] = E[XY] - \mu_X \underbrace{E[Y]}_{=\mu_Y} - \mu_Y \underbrace{E[X]}_{=\mu_X} + \mu_X\mu_Y$$

  $$\text{cov}(X,Y) = E[XY] - \mu_X\mu_Y = E[XY] - E[X]\cdot E[Y]$$ ← Note: If X,Y are Independent Then Cov(X,Y)=0

- **Correlation Coefficient ("Normalized and Centered Dependency")**

  $$\rho_{XY} = \frac{\text{cov}(XY)}{\sigma_X \cdot \sigma_Y} \; ; \; \sigma_X \equiv \sqrt{\text{var}(X)} \; ; \; \sigma_Y \equiv \sqrt{\text{var}(Y)} \qquad -1 \le \rho_{XY} \le 1$$

- **Standardized RV Transformation:**

  $$Z \equiv \frac{X-\mu_X}{\sigma_X} \qquad X: RV(\mu_X, \sigma_X^2) \implies Z: RV(\mu_Z = 0, \sigma_Z^2 = 1) \qquad \textit{zero mean \& unit variance}$$

A *single random variable* X is fully characterized by its probability mass function (PMF), $p_X(x)$, which in turn can be characterized *less precisely* by two numbers, namely, the mean or 1st moment $m_1 = \mu_X = E[X]$ and the 2nd moment, $m_2 = E[X^2]$. The 2nd moment is usually re-expressed as a *centered 2nd moment* which also goes by the names *variance* or *squared standard deviation*. The relation is obtained as follows:
$$\text{var}(X) = E[(X-\mu_X)^2] = E[(X^2 - 2\mu_X X + \mu_X^2] = E[X^2] - \mu_X^2$$
Common notation for the variance is squared standard deviation $\sigma_X^2$, which has equivalent useful forms
$$\sigma_X^2 = E[(X-\mu_X)^2] = E[X^2] - E[X]^2 = E[X^2] - \mu_X^2;$$
the square root of the variance or standard deviation $\sigma_X$ is most often quoted.
In a similar manner, a *pair of random variables* X and Y is fully characterized by their joint PMF, $p_{XY}(x,y)$, which in turn can be characterized, *less precisely*, by its two 1st moments yielding means $\{\mu_X, \mu_Y\}$ and a *symmetric matrix of 2nd moments* consisting of two variances $\{\text{var}(X), \text{var}(Y)\}$ and a single covariance $\text{cov}(X,Y)$ expressing their dependence. We note that for a joint PMF, $p_{XY}(x,y)$, the mean of X is by definition
$$E[X] = \sum_{x,y} p_{XY}(x,y)*x = \sum_x x\cdot(\sum_y p_{XY}(x,y)) = \sum_x x\cdot p_X(x) \; ; \; \text{with } p_X(x) = \text{marginal PMF on x}$$
With this understanding, the 1st moments of X and Y are single variable sums over their respective *marginal* PMFs and yield the two means:      $\mu_X = E[X]$ and $\mu_Y = E[Y]$.
The covariance matrix elements are defined by the three centered 2nd moments as follows:
$$\sigma_X^2 = \text{cov}(X,X) = \text{var}(X) = E[(X-\mu_X)^2] = E[X^2] - \mu_X^2$$
$$\sigma_Y^2 = \text{cov}(Y,Y) = \text{var}(Y) = E[(Y-\mu_Y)^2] = E[X^2] - \mu_X^2$$
$$\text{cov}(X,Y) = E[(X-\mu_X)(Y-\mu_Y)] = E[(XY-\mu_X Y - \mu_Y X + \mu_X\mu_Y] = E[XY] - \mu_X\mu_Y$$
The covariance is often "normalized" to give a correlation coefficient $\rho$, which is defined as
$$\rho_{X,Y} = \text{cov}(X,Y)/[(\text{var}(X)*\text{var}(Y)] = \text{cov}(X,Y)/[\sigma_X * \sigma_Y]$$
It can be shown that the correlation coefficient only takes on values in the interval $\rho_{X,Y} \varepsilon [-1,1]$.

## 6.6.1 Relationships for Derived RVs

# Relationships for Derived RVs

1. $E[aX + b] = a \cdot E[X] + b$
2. $E[aX + bY] = a \cdot E[X] + b \cdot E[Y]$
3. $E[X \cdot Y] = E[X] \cdot E[Y]$  *if  X, Y independent*
4. $\text{var}(X) = \text{cov}(X, X) = E[X^2] - E[X]^2$
5. $\text{var}(aX + b) = a^2 \cdot \text{var}(X)$
6. $\text{var}(aX + bY) = a^2 \cdot \text{var}(X) + b^2 \cdot \text{var}(Y) + 2ab \underbrace{\text{cov}(XY)}_{= \rho \cdot \sigma_X \sigma_Y}$

$$= a^2 \sigma_X^2 + b^2 \sigma_Y^2 + 2ab \cdot \rho \cdot \sigma_X \sigma_Y$$

$\theta = 90° \Rightarrow$ X & Y are independent variables yields **right triangle** →

Statistical Orthogonality :
If $\theta = 90°$ then
there is no correlation, *i.e.,* $\rho = 0$

7. $E[\sum_{k=1}^{n} X_k] = \sum_{k=1}^{n} E[X_k]$   *does not require independence*

8. $\text{var}(\sum_{k=1}^{n} X_k) = \sum_{k=1}^{n} \text{var}(X_k)$   *requires $X_k$ to be mutually independent for $k = 1, 2, ..., n$*

This slide states (without proof) a number of useful relationships resulting from linear transformations among one and two RVs.

Under the linear Y=aX+b and bi-linear Z=aX+bY transformations (1) and (2) show that the first moment characteristics transfer linearly. In (3), the expectation of the product of two RVs E[X*Y] equals the product of their individual expectations E[X]*E[Y] *only if* X and Y *are independent*. Under the linear transformation Y=aX+b, (5) shows that the 2nd moment quantity var(aX+b) is unaffected by the additive constant "b" and is proportional to the multiplier squared "$a^2$".

For two RVs X, Y under the bi-linear transformation Z=aX+bY, the 2nd moment quantity var(aX+bY) yields an expression equal to the separate multipliers squared times the individual variances, that is, $a^2$*var(X) + $b^2$*var(Y) plus a "cross-term" 2ab*cov(XY). If X and Y are independent this cross-term vanishes and the resulting variance satisfies var(Z)=var(aX+bY)= $a^2$*var(X) + $b^2$*var(Y), which gives a sum-of-squares right triangle relationship between the perpendicular "sides" $a\sigma_X$, $b\sigma_Y$, and the hypotenuse $\sigma_Z$. This is known as *"statistical orthogonality"* and expresses the *independence* of the two random variables X and Y.

Expression 7) shows that the expectation of the sum of many variables E[$\sum X_k$] is the sum of the individual expectations $\sum$E[$X_k$], an obvious generalization for the sum of two. On the other hand, expression 8) shows that the var($\sum X_k$ ) is generally not equal to $\sum$var($X_k$) unless all *variables are mutually independent*; this is the generalization of (6) to many variables. The n x n covariance matrix corresponding to all centered 2nd moments will be diagonal only if the RVs are *pair-wise independent* and this has an analogous interpretation of statistical orthorgonality, now in an n-dimensional space.

## 6.7 Conditional Statistics: "Wheel of Fortune"

# Conditional Statistics: "Wheel of Fortune"

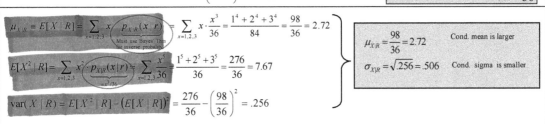

**a) Spin Wheel:**
Prob #1,#2,#3

$$p_X(x) = \begin{cases} \dfrac{x}{6} & x=1,2,3 \\ 0 & \text{otherwise} \end{cases}$$

**b) Draw from Numbered Container:**
Prob of Red "R" Marble
Given #1,#2,#3

$$p_{R|X}(r|x) = \begin{cases} \dfrac{x^2}{14} & x=1,2,3 \\ 0 & \text{otherwise} \end{cases}$$

#1: p=1/14    #2: p=4/14    #3: p=9/14

"Containers of Red & Blue Marbles"
"Similar to Red Trout in 3 Lakes"

Red Marble Outcomes

$p_{R|X}(r|x)$ : 1/14 R ← 1/84

$p_X(x)$ : 13/14 B

Start → 1/6 (1) 4/14 R ← 8/84 ; 10/14 B
2/6 (2)
3/6 (3) 9/14 R ← 27/84 ; 5/14 B

Compute Prob of Container X given Red drawn, *i.e.*, "Inverse Probability" $p_{X|R}(x|r)$

**a) Wheel: Container Choice Statistics**

$$\mu_X \equiv E[X] = \sum_{x=1,2,3} x \cdot p_X(x) = \sum_{x=1,2,3} x \cdot \frac{x}{6} = \frac{1^2+2^2+3^2}{6} = \frac{7}{3}$$

$$\boxed{\mu_X = \frac{7}{3} = 2.33}$$

$$E[X^2] = \sum_{x=1,2,3} x^2 \cdot p_X(x) = \sum_{x=1,2,3} \frac{x^3}{6} = \frac{36}{6} = 6 \qquad \text{var}(X) = E[X^2] - (E[X])^2 = 6 - \left(\frac{7}{3}\right)^2 = \frac{5}{9}$$

$$\boxed{\sigma_X = \sqrt{\frac{5}{9}} = .745}$$

**b) Need Inverse PMF $p_{X|R}(x|r)$** *(Conditional Statistics on X Given Red is Drawn)*

"Joint PMF" on X, R   $p_{RX}(r,x) = p_{R|X}(r|x) \cdot p_X(x) = \frac{x^2}{14} \cdot \frac{x}{6} = \frac{x^3}{84}$   Marginal PMF on R   $p_R(r) = \sum_{x=1,2,3} p_{R|X}(r,x) = \sum_{x=1,2,3} \frac{x^3}{84} = \frac{36}{84}$

"Joint PMF" on X, B   $p_{BX}(b,x) = p_{B|X}(b|x) \cdot p_X(x) = \left(1-\frac{x^2}{14}\right) \cdot \frac{x}{6} = \frac{14 \cdot x - x^3}{36}$   Inverse PMF   $\boxed{p_{X|R}(x|r) = \frac{p_{RX}(r,x)}{p_R(r)} = \frac{x^3/84}{36/84} = \frac{x^3}{36}}$

$$\mu_{X|R} \equiv E[X|R] = \sum_{x=1,2,3} x \cdot p_{X|R}(x|r) = \sum_{x=1,2,3} x \cdot \frac{x^3}{36} = \frac{1^4+2^4+3^4}{84} = \frac{98}{36} = 2.72$$
*Must use Bayes' Thm for inverse probability.*

$$E[X^2|R] = \sum_{x=1,2,3} x^2 \cdot p_{X|R}(x|r) = \sum_{x=1,2,3} \frac{x^5}{36} = \frac{1^5+2^5+3^5}{36} = \frac{276}{36} = 7.67$$
$_{= x^3/36}$

$$\text{var}(X|R) = E[X^2|R] - (E[X|R])^2 = \frac{276}{36} - \left(\frac{98}{36}\right)^2 = .256$$

$$\mu_{X|R} = \frac{98}{36} = 2.72 \qquad \text{Cond. mean is larger}$$
$$\sigma_{X|R} = \sqrt{.256} = .506 \qquad \text{Cond. sigma is smaller}$$

A joint distribution is generated by first spinning a "wheel of fortune" to give the bowl number X with probabilities $p_X(x) = x/6$ for x= 1,2,3, and then drawing a ball from that bowl with conditional probabilities $p_{R|X}(r|x) = x^2/14$ for red (R) and $p_{B|X}(b|x) = 1- x^2/14$ for blue(B) as illustrated in the figure. The tree shows the *a priori* probabilities for the bowls are {1/6, 2/6, 3/6}; the conditional branch probabilities from bowls #1, #2, #3 to red (R) are{1/14, 4/14, 9/14} and to blue (B) they are {13/14, 10/14, 5/14}. We see that there are three outcomes yielding R, and further that the probabilities along these three paths can easily be computed and summed to yield the total probability of red as P(R) =1/6(1/14) +2/6(4/14)+3/6(9/14) = 36/84 =.429. The inverse probabilities are conditioned on the outcome R and "look back" in the tree to find which bowl X the red ball R came from (same as Three Lakes Problem Slide#4-12); values of $p_{X|R}(x|r)$ are computed directly from the tree by simply taking ratios of the individual paths to the total, *viz.*,

$$p_{X|R}(x|r) = \{1/84, 8/84, 27/84\}/ (36/84) = \{1/36, 8/36, 27/36\}.$$

The PMF $p_X(x)$ is used to compute the 1st and 2nd moments of the distribution, which yields mean $\mu_X$= 2.33 and sigma $\sigma_X$ =.745. The conditional statistics for "X given R" are obtained using the above $p_{X|R}(x|r)$ as follows:

(i)     $\mu_{X|R}$ =E[X|R] = 1*(1/36) + 2*(8/36) + 3*(27/36) = 98/36= 2.72,

(ii)     $E[X^2|R] = 1^2*(1/36) + 2^2*(8/36) + 3^2*(27/36) = 98/36 = \{1+32+243\}/36 = 7.67$

(iii)     $\sigma_{X|R}^2 = E[X^2|R] - E[X|R]^2 = 7.67-(2.72)^2 = 0.272$

The inverse conditional probability $p_{X|R}( x|r)$ can also be computed analytically as follows:

(i)     compute the (R-component of) joint distribution as $p_{XR}(x,r) = p_{R|X}(r|x) p_X(x) = (x^2/14) * x/6 = x^3/84$,

(ii)     compute the marginal $p_R(r)$ (for color C = Red) by summing $p_{XR}(x,r)$ over x to yield
$p_R(r)= 1^3/84 + 2^3/84 + 3^3/84 = 36/84$     (same as total probability P(R) from the tree)

(iii)     compute the desired inverse conditional probability from its definition to yield
$p_{X|R}(x|r) = p_{XR}(x,r) / p_R(r) = (x^3/84) / (36/84) = x^3/36$ for x= 1,2,3 or {1/36, 8/36, 27/36}.

# 7 Properties of Random Variables

## 7.1 *Common PMFs and Properties -1*

### Common PMFs and Properties -1

| RV Name | PMF | Mean $E[X] = \sum_{x=0,1} x \cdot p_X(x)$ | Variance $\text{var}(X) = E[X^2] - E[X]^2$ |
|---|---|---|---|
| **Bernoulli** <br> *1-Trial* <br> *X=x succ.* <br> "0" or "1" *successes* | "Atomic" RV <br><br> $p_X(x) = \begin{cases} p & X=1 \text{ (success)} \\ 1-p=q & X=0 \text{ (failure)} \end{cases}$ | $E[X] = 0 \cdot (1-p) + 1 \cdot p$ <br> $= p$ | $E[X^2] = 0^2 \cdot (1-p) + 1^2 \cdot p$ <br> $= p$ <br> $\text{var}(X) = p - p^2 = p(1-p)$ <br> $= pq$ |
| **Binomial** <br> *n - Trials* <br> *X=x Succ.* <br> *How many succ "x" in "n" trials ?* | n Independent <br> **Bernoulli Trials** <br><br> $p_X(x) = \binom{n}{x} p^x q^{n-x}$ <br><br> $x = 0,1,\cdots n$ | $E[X] = \sum_{x=0}^{n} x \binom{n}{x} p^x q^{n-x}$ <br> $= np$ | $\text{var}(X) = npq$ |
| **Geometric** <br> *X=x Trials* <br> *1- Success* <br> How many trials "x" for "1" succ | **One Sequence** <br><br> $p_X(x) = \begin{cases} pq^{x-1} & x=1,2,\cdots \\ 0 & \text{(otherwise)} \end{cases}$ <br><br> Geom RV = Neg Binom <br> for r=1 succ. | $E[X] = \sum_{x=1}^{\infty} x \cdot pq^{x-1} = p\frac{d}{dq}\sum_{x=1}^{\infty} q^x$ <br> $= p\frac{d}{dq}\left(\frac{1}{1-q}\right) = \frac{+p}{(1-q)^2} = \frac{1}{p}$ <br> As $p$ decr. Expected num. trials <br> "x" for 1-succ must incr. | $\text{var}(X) = \frac{q}{p^2}$ |
| **Negative Binomial** <br> *X=x Trials* <br> *r- Successes* | **Many Sequences** <br> $p_X(x) = \underbrace{\binom{x-1}{r-1} p^{r-1} q^{x-r}}_{(r-1)\text{succ. in }(x-1)\text{ trials}} \cdot \underbrace{p}_{\substack{\text{succ. on} \\ \text{next trial}}}$ <br><br> $x = r, (r+1), (r+2), \cdots \infty$ | $E[X] = \sum_{x=r}^{\infty} x \binom{x-1}{r-1} p^r q^{x-r} = \frac{r}{p}$ <br> As $p$ decr. Expected num. trials <br> "x" for r-succ must incr. | $\text{var}(X) = r \cdot \frac{q}{p^2}$ |

This table and one to follow compare some common probability distributions, explore their fundamental properties and relate them to one another. In the table, a brief description is given under the "RV Name" column followed by the PMF formula and figure in col#2; formulas for the mean and variance are shown in the last two columns.

**The Bernoulli RV X** answers the question "what is the result of a single Bernoulli trial?" It takes on only two values, namely "1"=Success with probability p and "0"=Fail with probability q=1-p.

**The Binomial RV "X"** answers the question "how many successes X, are there in n Bernoulli trials?" It takes on values corresponding to the number of successes "X" in "n" independent Bernoulli trials; the sum variable, $X=X_1 + X_2 + \ldots + X_n$, gives the #successes by adding a 1 to the sum for each success only. The PMF $^nC_x \, p^x \, q^{n-x}$, is the binomial term whose coefficient is the #ways to have X=x successes in n trials; the single trial probability of success p enters as the product $p^x(1-p)^{n-x}$. A tree gives a nice visualization by explicitly showing the #tree-paths, $^nC_x$, that yield X=x successes at the tree output.

**The Geometric RV X answers the question** "how many Bernoulli trials X for 1 success?" It takes on values *from* 1 *to* $\infty$ and is the result of n-1 *failed Bernoulli trials* followed by one success; the sum of n Bernoulli RVs, $X=X_1 + X_2 + \ldots + X_n$ gives just "1" success and the tree has but a single path with X= x trials and only1-success yields the geometric PMF $q^{x-1}p^1$.

**The Negative Binomial RV X answers the question** "how many Bernoulli trials X for r- successes?" It takes on values *from* r *to* $\infty$ and is the *sum of* r *geometric RVs* $N_k$ with probability $p^{r-1}q^{nk-r}p^1$ that each count the #trials $N_k=n_k$ for its one success; the sum of r geometric RVs, $X=N_1+N_2+ \ldots +N_r$ gives "r" successes and the tree has $^{x-1}C_{r-1}$ paths for X=x-1 trials yielding (r-1)-successes, followed by one final success to yield the negative binomial PMF $^{x-1}C_{r-1} \, p^{r-1}q^{x-r} \cdot p^1 = {}^{x-1}C_{r-1} \, p^r q^{x-r}$ with x = r, r+1, ... $\infty$.

## 7.1.1 Bernoulli/Binomial Tree Structures

# Bernoulli/Binomial Tree Structures

| RV Name | PMF | Tree Graph |
|---|---|---|
| **Bernoulli** 1-*Trial* *X=x succ.* *"0" or "1"* *successes* | $p_X(x) = \begin{cases} p & X=1 \text{ (success)} \\ 1-p=q & X=0 \text{ (failure)} \end{cases}$  "Atomic" RV | **(q+p)** |
| **Binomial** *2 - Trials* *X=x Succ.* *How many succ "x" in "2" trials?* | $p_X(x) = \binom{2}{x} p^x q^{2-x}$  $x=0,1,2$  Independent Bernoulli Trials | **(q+p)²** |
|  | $(q+p)^2 = q^2 + 2pq + p^2$  $= {}^2C_0\,p^0\,q^2 + {}^2C_1\,p^1\,q^1 + {}^2C_2\,p^2\,q^0$ | |

The RVs of the last slide are grouped in pairs {Bernoulli, Binomial} and {Geometric, Negative Binomial} for a reason. The sum of many independent Bernoulli trials generates a Binomial distribution and similarly the sum of many independent Geometric trials generates the Negative Binomial distribution. This slide and the next give a graphical construction of trees for these two groups of paired distributions by repeatedly applying the basic tree structure of the underlying Bernoulli or Geometric tree structure to each output node of the previous stage.

In the first panel we show the PMF properties for Bernoulli on the left and on the right we display the Bernoulli tree structure where the upper branch q=Pr[Fail] goes to the state X= 0 and the lower branch p = Pr[Success] goes to the state X= 1.

In the second panel we show the PMF properties for a simple n=2 Binomial. The corresponding tree structure for this Binomial is obtained by appending a second Bernoulli tree to each output node of the first trial, thus yielding the 4 output states {{FF}, {FS}, {SF}, {SS}}. We see that there is $^2C_0$ tree paths leading to {FF} $p^0q^2$, $^2C_1$ tree paths leading to {FS} $p^1q^1$, and $^2C_2$ tree paths leading to {SS} $p^2q^0$, which is precisely as expected from the Binomial PMF for n=2.

This can be continued for n=3, 4, ... by repeatedly appending a binary Bernoulli tree to each new node. Furthermore, we see that this structure for n=2 is represented algebraically by $(q+p)^2$ since a direct expansion gives $1=q^2 + 2q^1p^1 +p^2$; expanding an expression corresponding to n Bernoulli trials $(q+p)^n$ obviously yields the appropriate Binomial expansion for the exponent n.

Thus the Binomial is represented by the repetitive tree structure or by the repeated multiplication of the algebraic structure 1=(q+p) by itself n-times to obtain $1^n=(q+p)^n$ .

## 7.1.2 Geometric/NegBinomial Tree Structures

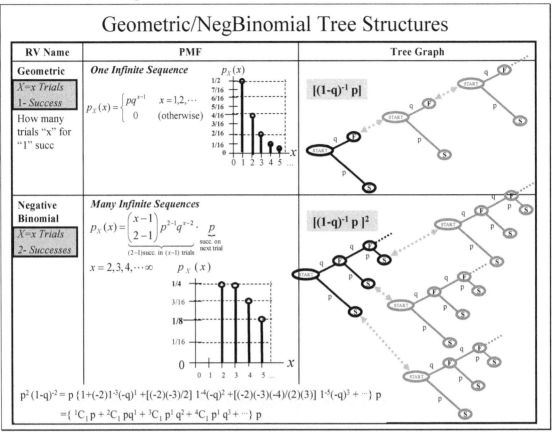

This slide first gives a graphical construction of a Geometric tree from an infinite number of Bernoulli trials and then shows how the Negative Binomial tree is the result of appending a Geometric tree to itself in a manner similar to that of the last slide. In the first panel we repeat the PMF properties for Geometric RV. On the right side of this panel we display Geometric tree structure whose branches end in a single success. This tree has a Bernoulli trial appended to *each failure node only* and is constructed from an infinite number of Bernoulli trials. The 1st Bernoulli trial yields X=1 with p=Pr[Success] and this ends the lower branch; its upper branch yields X=0 with q=Pr{Fail]; this failure node spawns a 2nd Bernoulli trial which again leads to X=1 or X=0; this process continues indefinitely. It accurately describes the probabilities for a single success in 1, 2, 3,...,∞ #trials and is algebraically represented by the expression $1=[(1-q)^{-1}p]$ which expands to give an infinite sequence $[1 + q^1 + q^2 + q^3 +....]\cdot p$, corresponding to exactly 0, 1, 2, 3,..."failures before a single success."

In the second panel we show the PMF properties for an r=2 Negative Binomial; on the right we display the Negative Binomial tree structure obtained by applying the basic Geometric tree only to each S-node (infinite number) corresponding to a 1st success. This leads to a doubly infinite tree structure for the r=2 Negative Binomial representing the number of trials X =x required for r=2 successes. We can verify the first few terms in the Negative binomial expansion given under PMF in the lower panel using the tree.

This process may be extended to r=3, 4, ... successes by repeatedly applying the Geometric tree to each success node. For n=2, direct expansion of the algebraic identity yields
$$1^2 = [(1-q)^{-1}p]^2 = \{ {}^1C_1 p + {}^2C_1 pq^1 + {}^3C_1 p^1 q^2 + {}^4C_1 p^1 q^3 + \cdots \}p,$$
which is in agreement with the n=2 Negative Binomial terms given in the table. In an analogous fashion, expansion of the algebraic generator $1^r = [(1-q)^{-1}p]^r$ for exponent r yields the correct result for the Negative Binomial. Note that the "Negative" modifier to Binomial is a natural designation in view of the $(1-q)^{-1}$ term in the algebraic structure.

## 7.1.3 Bernoulli, Geometric, Binomial and Negative Binomial PMFs

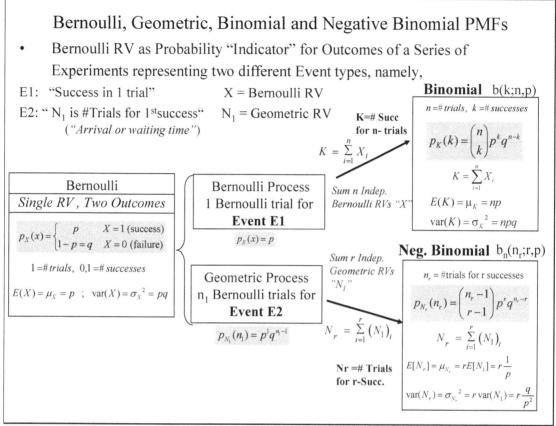

The Bernoulli RV is the basic building block for constructing other RVs; it has a PMF with only two outcomes, namely, a value X=1 with probability p and a value X=0 with probability q=1-p. We have seen that sum of n Bernoulli RVs yields a binomial RV with PMF b(x; n, p) for x=0, 1, 2,...,n , giving the *probability for "x" successes in "n" trials*. We have also seen that this can be understood by repeatedly appending the Bernoulli tree graph to itself, attaching this atomic binary tree to *all previous nodes*, thereby constructing a tree with $2^n$ outcomes corresponding to the n independent Bernoulli trials, each with two possible outcomes.

If instead, we want the *probability that "x" trials to produce "1" success*, we have essentially swapped the roles of *trials and successes*. In this case, we construct a Geometric PMF by repeatedly appending a Bernoulli tree graph to itself, but this time *only to the failure nodes*, thereby constructing a tree with an infinite number of output nodes, which correspond to "x-1" failures and exactly 1 success for trials x=1,2, ...., ∞.

Just as the Bernoulli tree graph is a building block for the Binomial tree graph, the infinite Geometric tree graph is a building block for the Negative Binomial. The Negative Binomial tree graph for r=2 successes is constructed by appending a Geometric tree graph to itself, *but now only to the previous success nodes*, resulting in a doubly infinite tree graph corresponding to exactly "x-1" failures and exactly 2 successes for x= 2,3 ...., ∞. Repeating this process r-times layer-by-layer yields the r-fold infinite tree graph corresponding to exactly "x-1" failures and exactly r successes for x= r,r+1, ...., ∞. The mathematical transformations relating Bernoulli, Binomial, Geometric, and Negative Binomial are developed on this slide.

## 7.2 *Common PMFs and Properties-2*

# Common PMFs and Properties-2

| RV Name | PMF | Mean $E[X]=\sum_{x=0,1}x\cdot p_X(x)$ | Variance $var(X)=E[X^2]-E[X]^2$ |
|---|---|---|---|
| **Hyper-geometric** $X=x$ -*succ* $N=$*fixed pop* $m=$*tagged* $n=$*test sampl* *w/o rplcemt* | $p_X(x)=\begin{cases}\dfrac{\overbrace{\binom{m}{x}}^{\substack{x\ from\\ "m-marked"}}\overbrace{\binom{N-m}{n-x}}^{\substack{(n-x)\ from\\ "(N-m)=\ unmarked"}}}{\binom{N}{n}}; & x_{min}\le x\le x_{max}\\[2ex] 0 & Otherwise\end{cases}$ $m,n\in[1,N];\;\; max(0,N-m-n)\le x\le min(m,n)$ PMF Derives from Binomial Identity $n\le m\le N$ $\binom{N}{n}=\binom{m+(N-m)}{n}=\binom{m}{0}\binom{N-m}{n}+\binom{m}{1}\binom{N-m}{n-1}+\cdots+\boxed{\binom{m}{x}\binom{N-m}{n-x}}+\cdots+\binom{m}{n}\binom{N-m}{0}$ | $E[X]=n\cdot\dfrac{m}{N}=n\cdot p$ where $p=m/N$ is the "initial" probability of drawing a marked item | $var(X)=\dfrac{(N-n)}{(N-1)}\cdot n\cdot\dfrac{m}{N}\cdot\dfrac{(N-m)}{N}$ $var(X)=\dfrac{(N-n)}{(N-1)}\cdot n\cdot p\cdot q$ |
| **Poisson** *#Trials* $\to\infty$ $X=x$ *Succ.* | $p_X(x)=\begin{cases}\dfrac{(a^x/x!)}{e^a} & x=0,1,2,\cdots\ \infty\\[1ex] 0 & Otherwise\end{cases}$ Limit of Binomial $a=\lim_{\substack{n\to\infty\\ p\to 0}}(n\cdot p)=\lambda\cdot t=$ (aver. arrival rate)*time | $E[X]=a$ | $var(X)=a$ |
| **Zeta(Zipf)** *#Trials* $\to\infty$ $X=x$ *Succ.* | **Riemann Zeta** $\zeta(s)=\sum_{x=1}^{\infty}1/x^s$ $\boxed{s>1}$ $p_X(x;s)=\begin{cases}\dfrac{1/x^s}{\zeta(s)}=\dfrac{"\zeta-term"}{\zeta(s)} & x=1,2,\cdots\\[1ex] 0 & (otherwise)\end{cases}$ | $E[X;s]=\dfrac{\zeta(s-1)}{\zeta(s)}$ $E[X;3.5]=\dfrac{\zeta(2.5)}{\zeta(3.5)}=1.191$ | $Var(X;s)=\dfrac{\zeta(s-2)}{\zeta(s)}-\left(\dfrac{\zeta(s-1)}{\zeta(s)}\right)^2$ $Var(X;3.5)=\dfrac{\zeta(1.5)}{\zeta(3.5)}-\left(\dfrac{\zeta(2.5)}{\zeta(3.5)}\right)^2$ $=.856$ |

This second Common PMFs table shows the hypergeometric, Poisson, and Riemann Zeta (or Zipf ) PMFs.

**The hypergeometric RV "X"** answers the question "how many successes (*defectives*) x are found in n test samples randomly selected from a production run containing m defective and N-m working items?" The Bernoulli trials here are **dependent** because the selections are made *without replacement*; the distribution is best understood by considering the binomial identity (see (6) on slide#2-22)

$$^{N}C_n={}^{m}C_0\ {}^{N-m}C_n+...+{}^{m}C_x\ {}^{N-m}C_{n-x}+...+{}^{m}C_m\ {}^{N-m}C_{n-m}$$

Dividing by $^{N}C_n$ yields a sum of terms equal to unity and thus defines the distribution term $^{m}C_x\ {}^{N-m}C_{n-x}/{}^{N}C_n$ where x only takes on values in the restricted range x=[$x_{min}$, $x_{max}$], with $x_{min}$=N-n-m and $x_{max}$= min(n,m). The table gives the mean as $\mu_X$ = np and the variance as var(X) = npq [(N-n)/(N-1)] where p = m/N is the 1st draw probability. Note that for large N these statistics are equivalent to those for a binomial b(x;n,p); but because the underlying Bernoulli trials are not independent the single trial probability of success p changes and there is a correction factor [(N-n)/(N-1)] for the hypergeometric distribution.

**The Poisson RV "X"** answers the question "how many successes x in n trials for n very large?" We shall discuss this in more detail in the second part of the course where we pair it with a continuous r-Erlang distribution. It is important to understand that the Poisson distribution represents a limiting behavior of the binomial PMF as n$\to\infty$ and that it is given by the ratio of a single Taylor expansion term of $e^a$ divided by $e^a$. *viz.*, $p_X(x)=\{a^x/x!\}/e^a$, for x=0,1,2,3,... The product a =$\lambda*$ t defines the so-called Poisson parameter, where $\lambda$ is the "rate of success" and t is the time interval over which data is taken. The Poisson RV has many applications in physics and engineering.

**The Riemann Zeta RV "X"** has applications to Language processing and prime number theory; a few of its properties are given in the table. Note that the exponent must satisfy s >1 in order to avoid the harmonic series whose sum does not converge and therefore cannot represent an actual PMF.

## 7.2.1 Statistical Analysis of Experimental Data

# Statistical Analysis of Experimental Data

**Bernoulli:** Single Trial
Two outcomes

$$p_X(x) = \begin{cases} p & X = 1 \text{ (success)} \\ 1 - p = q & X = 0 \text{ (failure)} \end{cases}$$

**Binomial:** n indep Bernoulli Trials
Each with two outcomes
p= single trial prob of success

$$p_X(x) \equiv b(x; n, p) = \binom{n}{x} p^x q^{n-x}$$

$$x = 0, 1, \cdots n$$

**Bernoulli = Binomial** for n=1:  $b(x; "1", p) = \binom{1}{x} p^x q^{1-x}$

**Experiment:** $N_B$ -independent *Bernoulli trials* (say $N_B$ =10,000)

Bern. Trial #   *1  2  3   4  5  6   7  8  9*

*1          2          3*

**3 "Bern. Trials/ Sample**

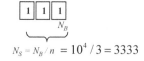

$$N_S = N_B / n = 10^4 / 3 = 3333$$

n=3
$\xi$ =#succ

| 2 | 1 | 0 | ••• | 3 |

Compare Expt'l frequency
of occurrence $f(\xi)$ in 3333
samples to theoretical PMF

Consider an "experiment" consisting of $N_B$=10,000 independent Bernoulli trials yielding data that is either a "0" or "1" in the 10,000 data bins as illustrated. The subsequent analysis of such data depends on precisely how the experiment was structured. For example, if all the data was taken in a single experiment "flipping a coin" repeated 10,000 times, then the correct analysis is to compare the results to a Bernoulli distribution with success probability p illustrated in the top figure. Thus, we would compute an experimental value of the Bernoulli success probability p by dividing the number of "1"s by 10,000; because we have so many samples we would feel pretty confident about this experimental value of $p_{exptl}$ = 2544/10000 =.2544 (say) and conclude that the underlying distribution is Bernoulli and that the *coin is unfair* (p≠.5).

Alternately, suppose that the data was taken in 3333 different sub-experiments in which three people simultaneously "flip a coin" and the data was recorded as a single concatenated sequence (with 1 bad 10,000[th] datum which we discard). In this case we would more naturally analyze the data in terms of 3333 samples (bottom figure) representing the sum of "group successes" which take on values {0,1,2, 3} corresponding to 0,1,2,3 heads. Thus, we would compare the experimental results with a Binomial distribution b(k; 3, p) by computing the frequency-of-occurrence for k=0,1,2,3 successes, this time dividing the number count for each by 3333. The experimental value of $p_{exptl}$ is not directly computable in this case since we now have 4 possible outcomes not just the 2 as in a single Bernoulli trial. We could estimate $p_{exptl}$ by adjusting p in the binomial until the theoretical probabilities most "closely match" the experimental results (frequencies-of-occurrence); but this requires a careful definition of "closely match" which is the subject of statistical testing. The main point here is that the same data can be analyzed differently depending on how it was actually collected.

## 7.2.1.1 Trade-Off: Sample Averaging vs. Number of Samples

# Trade-Off: Sample Averaging vs. Number of Samples

**Break up into arbitrary sample sizes n and the compare with appropriate binomial distribution with p=q=1/2**

---

**n=1: $N_S$ = $10^4$ samples**

    Compare sample frequency to Binomial b(x; **n=1**, p=1/2)=**Bernoulli** for x=0,1

    Compare sample mean to theoretical mean (**Bernoulli** ) E[X]= 1·p=1(1/2)=.5

    Compare sample Var to theoretical (**Bernoulli** ) Var(X) =1· pq=1(1/2)(1/2)=.25

*Too Granular*
*No Averaging*
*of "pixels"*

---

**n=3: $N_S$ =$10^4$/3 = 3333 samples**

    Compare sample frequency to Binomial b(x; **n=3**, p=1/2) for x=0,1,2,3

    Compare sample mean to theoretical (**sum 3 Bernoullis** ) E[X]=np=3(1/2)=1.5

    Compare sample Var to theoretical (**sum 3 Bernoullis** ) Var(X)=npq=3(1/2)(1/2)=.75

*Possible*
*Trade-off*
*strategy"*

---

**n=5000: $N_S$ = $10^4$/5000 = 2 samples**

    Compare sample frequency to Binomial b(x; **n=5000**, p=1/2) for x=0,...,5000

    Compare sample mean to theoretical (**sum 5000 Bernoullis** ) E[X]=np=5000(1/2)=2500

    Compare sample Var to theoretical (**sum 5000 Bernoullis** ) Var(X)=npq=5000(1/2)(1/2)=1250

*Poor Statistics*
*only 2 samples!*

---

**Must trade-off sample averaging *versus* number of samples**

Another important aspect of data analysis is to consider the trade-off between "sample averaging" which removes unwanted "experimental noise" and the *reduction in the total number of samples leading to less statistical significance of the result*. We shall consider grouping the data into sums of n samples (equivalent to averaging without dividing by n) and compare the sample mean and variance with the theoretical values computed from an assumed Binomial distribution b(x; n, p). (Below, we also assume p=q=1/2.)

**For n=1 (no averaging),** there are Ns = 10,000 samples, and Binomial b(x; 1, p=1/2) with x=0,1 (Bernoulli) and we compare against the theoretical values E[X]= n*p = 1*.5=.5 and var(X)=npq=1*.5*(1-.5)=.25 . The 10,000 samples yield good "statistical significance" but are subject to random noise variations that can distort the underlying data; thus the sample variance should be good but the sample mean may be corrupted by the noise and yield a poor estimate.

**For n=3 (3-sample averaging),** there are Ns = 3333 samples (1 is dropped), and Binomial b(x; 3, p=1/2) with x=0,1,2,3 and we compare against the theoretical values E[X]= n*p = 3*.5=1.5 and var(X) = npq = 3*.5*(1-.5) =.75. The 3333 samples still yield good "statistical significance" and now the random noise variations have been reduced by averaging 3 samples at a time; thus both the sample mean and sample variance should be "good" estimates.

**For n=5000 (5000-sample averaging),** there are now just Ns = 2 samples, and Binomial b(x; 5000,p=1/2) with x=0,1,2,3, ...,5000 and we compare against the theoretical values E[X]= n*p = 5000*.5=2500 and var(X)=npq=5000*.5*(1-.5)= 1250. The 2 samples have virtually no "statistical significance" and although averaging almost completely eliminates the random noise variations, we only have two samples and we do not expect the resulting estimates to be meaningful.

## 7.2.1.2 Statistical Analysis of N=16 Bernoulli Trials

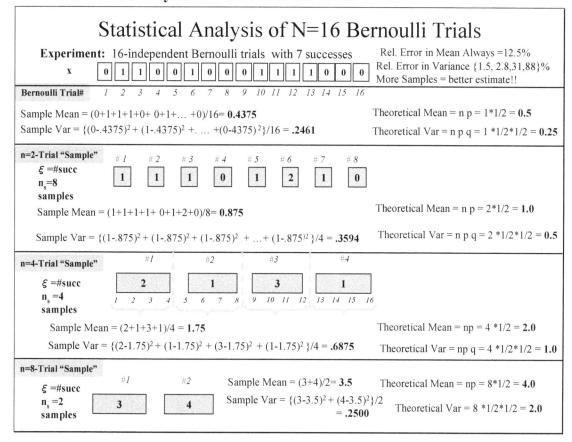

# Statistical Analysis of N=16 Bernoulli Trials

**Experiment:** 16-independent Bernoulli trials with 7 successes

Rel. Error in Mean Always =12.5%
Rel. Error in Variance {1.5, 2.8,31,88}%
More Samples = better estimate!!

x | 0 | 1 | 1 | 0 | 0 | 1 | 0 | 0 | 0 | 1 | 1 | 1 | 1 | 0 | 0 | 0

Bernoulli Trial# : 1 2 3 4 5 6 7 8 9 10 11 12 13 14 15 16

Sample Mean = (0+1+1+1+0+ 0+1+1+... +0)/16= **0.4375**

Theoretical Mean = n p = 1*1/2 = **0.5**

Sample Var = {(0-.4375)$^2$ + (1-.4375)$^2$ +. ... +(0-.4375)$^2$}/16 = **.2461**

Theoretical Var = n p q = 1 *1/2*1/2 = **0.25**

**n=2-Trial "Sample"**
$\xi$ =#succ
$n_s$=8
samples

# 1 | # 2 | # 3 | # 4 | # 5 | # 6 | # 7 | # 8
1 | 1 | 1 | 0 | 1 | 2 | 1 | 0

Sample Mean = (1+1+1+1+ 0+1+2+0)/8= **0.875**

Theoretical Mean = n p = 2*1/2 = **1.0**

Sample Var = {(1-.875)$^2$ + (1-.875)$^2$ + (1-.875)$^2$ + ...+ (1-.875)$^{2}$ }/4 = **.3594**

Theoretical Var = n p q = 2 *1/2*1/2 = **0.5**

**n=4-Trial "Sample"**
$\xi$ =#succ
$n_s$ =4
samples

#1 | #2 | #3 | #4
2 | 1 | 3 | 1

1 2 3 4 5 6 7 8 9 10 11 12 13 14 15 16

Sample Mean = (2+1+3+1)/4 = **1.75**

Theoretical Mean = np = 4 *1/2 = **2.0**

Sample Var = {(2-1.75)$^2$ + (1-1.75)$^2$ + (3-1.75)$^2$ + (1-1.75)$^2$ }/4 = **.6875**

Theoretical Var = np q = 4 *1/2*1/2 = **1.0**

**n=8-Trial "Sample"**
$\xi$ =#succ
$n_s$ =2
samples

#1 | #2
3 | 4

Sample Mean = (3+4)/2= **3.5**

Sample Var = {(3-3.5)$^2$ + (4-3.5)$^2$}/2 = **.2500**

Theoretical Mean = np = 8*1/2 = **4.0**

Theoretical Var = 8 *1/2*1/2 = **2.0**

Here is an explicit trade-off analysis for averaging n= 1, 2, 4, and 8 data samples given the data from 16 Bernoulli trials shown. For each case, we compute the experimental sample mean and variance and compare them with those for the assumed underlying Binomial b(x; n, 0.5) distribution.

**For n=1** we have 16 Bernoulli trials and compare *sample with theoretical statistics* to find: means .4375 *vs.* 0.5, and variances .2461 *vs.* 0.25. This clearly shows reasonably good results for both. (Of course we have not added noise to this data, so we will not see the effects of averaging.)

**For n=2** we have 8 samples and compare *sample with theoretical statistics* to find: means .875 *vs.* 1.0, and variances .3594 *vs.* 0.5. This clearly shows some degradation in the results for both. (However, there will be an improvement in noise reduction.)

**For n=4** we have 4 samples and compare *sample with theoretical statistics* to find: means 1.75 *vs.* 2.0, and variances .6875 *vs.* 1.0. This clearly shows increased degradation the results for both. (However, the improvement in noise reduction will be correspondingly greater.)

**For n=8** we have 2 samples and compare *sample with theoretical statistics* to find: means 3.5 *vs.* 4.0, and variances .2500 *vs.* 2.0. With only 2 samples not much is expected in either comparison. (However, this would have the best improvement in noise reduction.)

Finally, we note that the relative error in the mean is 12.5% in all cases, while the error in the relative variance increases as we average more data samples proceeding down the slide to yield 1.8%, 2.8%, 31%, and 88% errors. Thus, the smaller number of "averaged samples" proceeding down the slide yields progressively worse estimates of the statistical variability of the actual data as captured by its computed variance. It should be pointed out that since noise is usually distributed equally about the mean, it will have only a small effect on the computed mean, but a significant effect on the variance which measures squared deviations from the mean.

## 7.2.2 Bernoulli/Binomial Examples

# Bernoulli/Binomial Examples

*#Succ "x" for "n" Trials*

- **Bernoulli "Atomic" RV**
  - "1" trial , prob. of success p
  - "2" outcomes: 0= fail, or 1= success
  - Fair coin flipped once
  - H="1"  $P\{X=0\}=q=1-p$
  - T="0"  $P\{X=1\}=p$ ;

- **Binomial "Composite" RV**
  - "n" Bernoulli trials
  - Binomial Expansion: $(p+q)^n = 1^n = 1$
  - Tree of outcomes: $2^n$ permutations
  - Groups: $^2\mathscr{C}_n = {}^{2+n-1}C_n = {}^{n+1}C_n = {}^{n+1}C_1 = n+1$

### 5 Fair coins flipped

n = 5 Bernoulli trials

Single trial prob succ. p= ½

$2^5$=32 permutation outcomes

5+1=6 Groups:

$P\{X=0\} = {}^5C_0\,(\frac{1}{2})^0(1-\frac{1}{2})^5 = 1/32$

$P\{X=1\} = {}^5C_1\,(\frac{1}{2})^1(1-\frac{1}{2})^4 = 5/32$

$P\{X=2\} = {}^5C_2\,(\frac{1}{2})^2(1-\frac{1}{2})^3 = 10/32$

$P\{X=3\} = {}^5C_3\,(\frac{1}{2})^3(1-\frac{1}{2})^2 = 10/32$

$P\{X=4\} = {}^5C_4\,(\frac{1}{2})^4(1-\frac{1}{2})^1 = 5/32$

$P\{X=5\} = {}^5C_5\,(\frac{1}{2})^5(1-\frac{1}{2})^0 = 1/32$

### Defective screws

a) Sold in packages of n = 10

single trial prob fail = p=.01 , q=.99.

*Failure of at most one screw in a package of 10*

Replace if "more than" k=1 one screw is defective.

Find probability of replacement

Binomial b(x;p,n)=b(x;.01,10)

$P(X > 1) = 1 - P(X=0) - P(X=1)$

$= 1 - {}^{10}C_0\,(.01)^0(.99)^{10} - {}^{10}C_1\,(.01)^1(.99)^9 = .004$

b) 100 packages of 10  Y=10-packs

single trial prob fail = $p_{pkg}$ =.004 , q=.996.

*Failure of k packages out of 100*

$P(Y=k) = {}^{100}C_k\,(.004)^k\,(.996)^{100-k}$  **exactly k 10-packs fail**

The Bernoulli RV "X" is the basic building block for other RVs and has a PMF with only two outcomes X=1 with probability p, and X=0 with probability q = 1 - p. The sum of n independent Bernoulli random variables yields a binomial RV with PMF {b(x; n, p), x=0,1,2,...,n} giving the probability for "x successes in n trials." We have also seen that repeatedly appending the Bernoulli tree graph to itself *at each of its nodes* yields a binomial tree with $2^n$ outcomes.

**The binomial PMF answers the question "how many successes "x" in "n" trials?"**

In the first example, 5 fair coins are flipped simultaneously resulting in a binomial PMF with n=5 Bernoulli trials and single trial success probability p=1/2, *viz.*, b(x; n=5,p=1/2) = $^5C_x\,p^x q^{5-x}$ for x=0,1,2,3,4,5; the PMF values are calculated and displayed on the slide.

The **defective screw** example involves packages of hardware containing 10 screws and a guarantee for *package replacement* if "more than k=1 screw" is defective. Each package consists of 10 Bernoulli samples from a manufacturing lot with 1% failure probability p=.01 and q=1-p=.99. The probability of having X defective screws in a package of 10 is given by a Binomial PMF with the stated parameters as b(x; 10, .01) = $^{10}C_x\,p^x q^{10-x}$. In terms of this distribution, the probability that a package must be replaced is simply the probability of more than one failure, *viz.*,

$$P(X>1) = 1-\{P(X=0)+P(X=1)\}=1- {}^{10}C_0\,p^0 q^{10-0} - {}^{10}C_1\,p^1 q^{10-1} = .004 \text{ or } 0.4\%.$$

Changing our test unit to Y="10-packs", the above result becomes the single trial probability of a "10-pack" failure $p_{pkg}$ = .004. Now, testing n= 100 units (10-packs), the probability of exactly "k" failures is given by the binomial term P(Y=k) = $^{100}C_k(.004)^k\,(.996)^{100-k}$.

## 7.2.3 Geometric RV (Draw with Replacement)

# Geometric RV (Draw with Replacement)

*#Trials "x" for "1" Succ.*

- Urn with "N" White (W) and "M" Black (B) balls
  - Many independent trials (*w/replacement*) until B is drawn
  - **Single trial prob succ.** $p = M/(M+N)$ ; $q = (1-p) = N/(M+N)$
  - "2" outcomes: W = fail, B = success

- a) Find prob exactly "*k* " trials needed "means"

  Failure on $1^{st}$ (k-1)-trials

  Success on $k^{th}$ trial $\quad P(X = k) = q^{k-1} p^1 = \left(\frac{N}{N+M}\right)^{k-1} \left(\frac{M}{N+M}\right)$

  $\underbrace{\quad}_{(k-1)\text{ fail}} \underbrace{\quad}_{1\text{ succ.}}$

- b) Find prob **at least k+1 trials** needed, *i.e., k+1, k+2, ...*

  $$P(X > k) = \underbrace{q^k p^1}_{(k+1)\text{ trials}} + \underbrace{q^{k+1} p^1}_{(k+2)\text{ trials}} + \underbrace{q^{k+2} p^1}_{(k+3)\text{ trials}} + \cdots$$

  $$= q^k \cdot p\left(1 + q + q^2 + q^3 + \cdots\right) = q^k \cdot \frac{p}{(1-q)} = q^k$$

*Geometric RV*

$(1-q)^{-1} p$

k prior failures

$q^k$

$(k+2)^{nd}$ -trial

$(k+1)^{st}$ trial

$q^k \{(1-q)^{-1} p\}$

**Renewal Process**

CDF $\quad P(X \le k) \quad 1^{st}\text{ Succ} \quad 2^{nd}\text{ Succ}$

$P(X \le k) = 1 - q^k$
for $k = 1,2,3,\cdots$

failures    failures    failures

re-start    re-start

k=1    k=1    k=1
No memory of prior failures!!

**The Geometric PMF answers the question "how many trials "x" are needed for "1" success?"**
Given an urn containing N white (W) and M black (B) balls, one ball is drawn and then placed back into the urn until a black ball is drawn. Because of replacement, the number of W and B balls in the urn is always the same; thus each draw is *independent* and has the same "single trial probability of success (selecting B)," namely, $p=M/(M+N)$ [and $q=1-p= N/(M+N)$.] The answer to the question *"how many trials "x" are needed for exactly "1" success"* is characterized by a Geometric RV "X" with PMF, $p_X(x) = q^{x-1} p^1$, with $x=1,2,3,...$denoting the number of trials needed for a single success.
a) **Find probability "exactly k trials"** are needed: $P(X=k)= p_X(x=k) = (N/(M+N))^{k-1} (M/(M+N))$
b) **Find probability "at least k+1 trials"** are needed: $P(X>k) = q^k p^1 + q^{k+1} p^1 + q^{k+2} p^1 + q^{k+3} p^1 +...$ , which factors to
$$P(X>k) = q^k p^1 (1 + q^1 + q^2 + q^3 +...) = q^k p / (1-q) = q^k p / p = q^k$$
Note that "greater than k trials" requires a sum over an infinite number of possible paths to success, starting with k-failures prior to success on trial k+1, or (k+1)-failures prior to success on trial k+2, *ad infinitum*. The factoring calculation above shows that we have the same infinite series, even after k prior failures, so the result is *independent of prior failures*. The infinite sum always yields the same value 1/[1-q] which cancels the "p" term in the numerator leaving $q^k$ as the resulting probability.
This can be visualized as appending an entire geometric tree $\{(1-q)^{-1} p\}$ to the tree representing the first k-failures as illustrated in the slide. This means that the probability of *k more failures* is still $q^k$ and not $q^{2(k)}$; this is called a "renewal process" since it has no memory of the previous k-failures.
A plot of the cumulative distribution function (CDF) reveals the nature of this renewal process more intuitively. Thus setting $p=q=0.5$, the (CDF) can be written explicitly as
$$F(k) \equiv P(X \le k) = 1 - P(X>k) = 1 - (q)^k = 1 - (.5)^k$$
The CDF curve $F(k) = 1 - q^k = 1 - (.5)^k$ starts at .5 for k=1 and increases to .75 for k=2, .875 for k=3, ... and asymptotically approaches 1 as $k \to \infty$ as shown in the *first segment* of the plot. However, if a success occurs at say k=15, the whole process re-starts after this $1^{st}$ success by resetting the failures to zero and starting again with F(1)=.5 at k=1 as shown in the *second plot segment*; now the $2^{nd}$ success may occur after, say 100 failures, and the process restarts once more at k=1 and F(1)=.5, *ad infinitum*.

## 7.2.4 Banach Match Problem: Negative Binomial

# Banach Match Problem: Negative Binomial  *#Trials "x" for "r" Succ.*

- Banach Match Problem
  - Pipe-Smoking Mathematician – 2 boxes N – matches each
  - Left & Right pockets – Eq. Likely Random draw
  - **Discovers Left is Empty** - Event "E"
  - Find Prob Right has exactly "k" matches left
  - **"Succ"= Choose Left Match Box (prob** $p$**)**
  - Choose Left: r = N+1 times & "discover empty"

    in  x= (N+1)+(N-k) draws (k remain in Rt box)

$$p_X(x) = \underbrace{\binom{x-1}{r-1} p^{r-1} q^{x-r}}_{(r-1)\text{succ. in } (x-1)\text{ trials}} \cdot \underbrace{p}_{\substack{\text{succ. on}\\ \text{next trial}}}$$

$$x = r, (r+1), (r+2), \cdots \infty$$

$$p_X(X = "E") = \underbrace{\binom{(2N-k+1)-1}{(N+1)-1} p^N q^{(2N-k)-N}}_{(r-1)\text{succ. in } (x-1)\text{ trials}} \cdot \underbrace{p^1}_{\substack{\text{succ. on}\\ \text{next trial}}}$$

$$= \binom{2N-k}{N}\left(\frac{1}{2}\right)^{N+1}\left(\frac{1}{2}\right)^{(N-k)} = \binom{2N-k}{N}\left(\frac{1}{2}\right)^{2N-k+1}$$

*Geom RV = Neg Binom  for r=1 succ.*

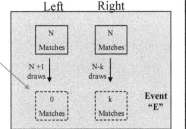

*Empty box could be either L or R, so multiply  by 2*

$$P(\text{Remain} = k) = 2p_X(x) = \binom{2N-k}{N}\left(\frac{1}{2}\right)^{2N-k}$$

**The negative binomial PMF answers "how many trials "x" are needed for "r" successes?"**
A pipe smoking mathematician has two identical boxes of N matches in his left and right pockets.  He randomly selects a match from either the left pocket with probability p or from the right pocket with probability q and continues this until he discovers that one box is empty.  Let E be the event that the left pocket matchbox is empty; find the probability that he has exactly k matches remaining in the right pocket matchbox.

We define a "success" to be a *draw from the left box* with single trial probability of success p.  The grey box figure illustrates the situation: N+1 draws are made from the Left matchbox leaving it empty while N-k draws are made from the right matchbox leaving it with k matches.  The negative binomial is used to compute the probability of the event E "left is empty" as follows: the total number of trials, x, is the sum of (N+1) from the left and (N-k) from the right or x=2N-k+1; the *number of successes* (draws from the left box) is r =N+1, so the answer P(E) is simply the negative binomial distribution term for x=2N-k+1 *trials* and r=N+1 *successes*.  Substituting these values we find

$$P(E) = p_X(x=2N-k+1; r=N+1) = \{ {}^{2N-k}C_N\, p^N\, q^{N-k} \} * p;$$

since this could equally well have been the right pocket matchbox we must double the result to obtain

$$P(k\text{-remain}) = 2\, P(E)$$

We note that if the trials are considered to take place at uniform time or "sampling" intervals, then the number of trials x, for r-successes is simply the time of the $r^{th}$ success, which is often denoted as the $r^{th}$ *arrival time*.  This gives another important aspect of probability calculations; not only are we interested in how many successes there are, but also we are often interested in the timing of the events denoted by their "arrivals" in time.

## 7.2.5 Banach Match Problem: Binomial RV

# Banach Match Problem: Binomial RV

**#Succ "x" in "2N-k" trials**

- Banach Match Problem Revisited
  - Pipe-Smoking Mathematician – 2 boxes with N – matches each
  - Left & Right pockets – Eq. Likely Random draw
  - **"Succ"= Choose Left Match Box (prob $p$)**
  - Replace L with p and R with q to obtain
  - Binomial PMF
    with $n = 2N-k$  $$(p+q)^{2N-k} = \sum_{\alpha=0}^{2N-k} \binom{2N-k}{\alpha} p^{\alpha} q^{(2N-k)-\alpha}$$

$$(L+R)^{2N-k} = \sum_{\alpha=0}^{2N-k} \binom{2N-k}{\alpha} L^{\alpha} R^{(2N-k)-\alpha}$$

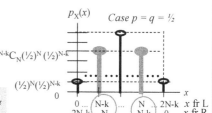

**Two equal binomial terms** of interest (N from L & N from R)

$$P(\text{N from Left}) = P(NL) = \binom{2N-k}{N}\left(\frac{1}{2}\right)^{2N-k}$$

$$P(\text{N from Right}) = P(NR) = \binom{2N-k}{N}\left(\frac{1}{2}\right)^{2N-k}$$

**Event E:** Either L or R is empty and other has k matches

$$P(E) = P(E \mid NR) \cdot P(NR) + P(E \mid NL) \cdot P(NL)$$

**Next draw from either L or R** yields event E with equal probability

$$P(E|NR) = P(E|NL) = 1/2$$

$$P(E) = \frac{1}{2} \cdot \binom{2N-k}{N}\left(\frac{1}{2}\right)^{2N-k} + \frac{1}{2} \cdot \binom{2N-k}{N}\left(\frac{1}{2}\right)^{2N-k} = \binom{2N-k}{N}\left(\frac{1}{2}\right)^{2N-k}$$

Alternately, the Banach Match Problem can be solved using the binomial distribution for matches drawn from the left (L) and right (R) pockets as $(L+R)^{2N-k}$, or in terms of their probabilities $(p+q)^{2N-k}$. The terms of this binomial expansion represent all the ways the 2N-k draws can be distributed between the L and R pockets. The figure shows this binomial b(x; 2N-k, p) for p = q = 1/2 and calls out two specific terms (red, bold), namely,
(i) $P(NL) = {}^{2N-k}C_N \, p^N \, q^{N-k}$ for *N draws from L (N-k draws from R)*, and
(ii) $P(NR) = {}^{2N-k}C_{N-k} \, p^{N-k} \, q^N$, for *N-k draws from L (N draws from R)*.
The event E that a matchbox is *discovered empty* is established by attempting to draw from an empty match box. Thus to discover L is empty given "NL" we attempt to draw one more from it with probability P(E|NL)=p; similarly to discover R is empty given "NR" we attempt drawing from it with probability P(E|NR)=q; thus writing total probability for the event empty (E) as the sum these two contributions we have
$$P(E) = P(E|NL)*P(NL) + P(E|NR)*P(NR), = p*\{{}^{2N-k}C_N \, p^N \, q^{N-k}\} + q*\{{}^{2N-k}C_{N-k} \, p^{N-k} \, q^N\}$$
The fact that this problem can be solved using arguments involving two entirely different distributions is curious and indicates that the solution is not a RV corresponding to either distribution. In both cases we use arguments relating to half the problem, *e.g.*, L matchbox and then add the symmetric result for the R matchbox. The Negative Binomial RV "$X_L$" represents the #draws x for r=N+1 successes (success = draw from the L), while the Negative Binomial RV "$X_R$" represents the #draws x for r=N+1 successes (success = draw from R) and although both "$X_L$" and "$X_R$" are Negative Binomial RVs their sum "$X_L + X_R$" is not. In fact, they are samples from two entirely different "mirror" sample spaces: L empty after "$X_L$ draws," and R empty after "$X_R$ draws." In the binomial case, we take two terms from the PMF b(x; **2N-k**, p), but then on the next draw the index "n" increases to n+1 ={2N-k} +1, so that the result P(E) is a realization of a different binomial distribution, *i.e.*, b(x; **2N-k+1**, p). In any case, both methods are useful and give insight into the problem.

## 7.2.6  Hypergeometric RV:  Lot Rejection

# Hypergeometric RV:  Lot Rejection

*Pop. "N" & Sub-Pop "m" (defectives)*
*#Succ. (defectives) "x" in "n" trials*

**Mfg. Lot Rejection**   *Two Lots of size N=10.  Inspect n= 3 and accept if none (x=0) are defective.*
Lots taken from a mixture of Two types: 30% Type1 with m=4 defective &
70% Type2 with m=1 defective.  Find probability of rejection

x from     (n-x) from
"m-tagged"  "N-m untagged"

m = tagged = defective population
N-m = untagged = working population

Choose " x"  from tagged population  "m"
Choose "n-x" from remaining pop. "N-m"

$$p_X(x; N, m, n) = \frac{\binom{m}{x}\binom{N-m}{n-x}}{\binom{N}{n}}$$

"n" from
Population "N"

First compute probability of Acceptance from two lots taken from the mixture:

$$P(Accept) = P(A \mid Type_1)\underbrace{P(Type_1)}_{=30\%} + P(A \mid Type_2)\underbrace{P(Type_2)}_{=70\%}$$

Accept(A ) means  x=0 defective; Lot Size: N=10;

Allowed defectives:  x=0 ;  Random Test Sample: n=3;

| Type$_1$ : m=4 defective | Type$_2$ : m=1 defective |
|---|---|
| x from     (n-x) from  "m-def"    "N-m def"  $$P(A \mid Type_1) = \frac{\binom{m}{x}\binom{N-m}{n-x}}{\binom{N}{n}} = \frac{\binom{4}{0}\binom{10-4}{3-0}}{\binom{10}{3}}$$  "n" from Population "N"   x=0;n=3; m=4;N=10 | x from     (n-x) from  "m-def"    "N-m def"  $$P(A \mid Type_2) = \frac{\binom{m}{x}\binom{N-m}{n-x}}{\binom{N}{n}} = \frac{\binom{1}{0}\binom{10-1}{3-0}}{\binom{10}{3}}$$  "n" from Population "N"   x=0;n=3; m=1;N=10 |

$$P(Accept) = \frac{\binom{4}{0}\binom{10-4}{3-0}}{\binom{10}{3}} \cdot 30\% + \frac{\binom{1}{0}\binom{10-1}{3-0}}{\binom{10}{3}} \cdot 70\% = 54\%$$         $P(Reject) = 1 - P(Accept) = 46\%$

**The hypergeometric PMF answers the question "how many successes "x" are drawn from a sub-population of "m" (defectives), when a total of n samples are taken from the general population N?"**
This has a rather simple interpretation in terms of testing a small number of samples "n" from a large production run of size "N. "  The PMF gives the probability of drawing (hence detecting) x "defectives" when it is known that the entire production run has "m" defectives and "N-m" non-defectives.  The PMF is constructed by taking *x from m defectives*,  taking *n-x from N-m non-defectives*, and then dividing their product by *total number of ways* of choosing the n test samples from the entire production run N, *viz.*,

$$p_X(x) = [{}^mC_x \ast {}^{N-m}C_{n-x}] / {}^NC_n$$

The example considers two lots of size N=10 and takes a random test sample of size n =3; the acceptance criterion is zero defectives x=0.  If the order is filled by randomly selecting 10 samples from two different types, then we must first compute the probability of acceptance (A) for each type separately since they have different hypergeometric distributions corresponding to m=4 for type$_1$ and m=1 for type$_2$.  Thus we find

$$P(A|type_1) = [{}^4C_0 \ast {}^{10-4}C_{3-0}] / {}^{10}C_3 = .167 \qquad \text{m=4 defective}$$
$$P(A|type_2) = [{}^1C_0 \ast {}^{10-1}C_{3-0}] / {}^{10}C_3 = .7 \qquad \text{m=1 defective}$$

The problem states that 30% of the test samples are from type$_1$ and 70% are from type$_2$ so P(type$_1$) = 30% and P(type$_2$) = 70% .  Substitution of these values into the total probability of acceptance expression we find

$$P(Accept) = P(A| type_1)\ast P(type_1) + P(A| type_2)\ast P(type_2) \ = .167\ast.30 +.7\ast.70 =.050 +.49 = .54$$

Thus the probability of rejection is P(Reject) = 1-.54 = .46 which is of course way too high.  Note that selecting all from the better lot still yields 1-.7=.30 which is better, but still not very good.
Another example is the PMF for the number of aces X resulting from n=20 draws from a deck of N=52 cards having m=4 aces and N-m=48 *non-aces*; direct substitution yields p$_X$(x), for X={0,1,2,3,4} aces

$$p_X(x) = \{{}^4C_0\,{}^{48}C_{20},\ {}^4C_1\,{}^{48}C_{19},\ {}^4C_2\,{}^{48}C_{18},\ {}^4C_3\,{}^{48}C_{17},\ {}^4C_4\,{}^{48}C_{16}\}/{}^{52}C_{20} = \{.132,.366,.348,.135,.018\}$$

## 7.2.6.1 Hypergeometric RV: Animal Population

# Hypergeometric RV: Animal Population

*Pop. "N" & Sub-Pop "m" (tagged)*
*#Succ. (tagged) "x" in "n" trials*

- **Unknown Animal Population "N"**
  - Catch# 1: m Tag & Release
  - Catch #2: n samples count x= #tagged
  - Maximum Likelihood Estim. of Pop. N
- Other Applications: Mfg Lot Rejection (previous slide), 21$^{th}$ card is an Ace, Urns, ...

$p_X(x; N) =$ prob of x observed in 2$^{nd}$ catch given pop N
Estimate of pop N is that value that maximizes $p_X(x; N)$

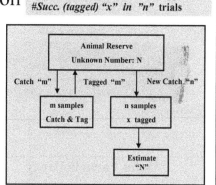

Ratio of successive terms as function of "N"

$$\frac{p_X(x;N)}{p_X(x;(N-1))} = \frac{\binom{m}{x}\binom{N-m}{n-x}\binom{N}{n}^{-1}}{\binom{m}{x}\binom{(N-1)-m}{n-x}\binom{(N-1)}{n}^{-1}}$$

| Case | N | p(x;N) |
|---|---|---|
| m=10 | 73 | .32392 |
| n=15 | 74 | .32409 |
| x=2 | 75 | .32409 |
|  | 76 | .32393 |

$$= \frac{(N-m)!}{(N-m-n+x)!(n-x)!} \cdot \frac{(N-m-n+x-1)!(n-x)!}{(N-m-1)!} \cdot \frac{(N-1)!}{n!(N-n-1)!} \cdot \frac{n!(N-n)!}{N!}$$

$$= \frac{(N-m)(N-n)}{N(N-m-n+x)} = 1 \qquad \Rightarrow \quad (N-m)(N-n) = N(N-m-n+x)$$

$$\Rightarrow N = \frac{mn}{x} \quad \text{... or a direct arguments} \rightarrow$$

(i) $\quad \dfrac{m}{N} \cong \dfrac{x}{n}$
frac of Pop Tagged 1$^{st}$ catch $\quad$ frac of tagged Pop in 2$^{nd}$ catch

(ii) $\quad \mu_X = E[X] = n\,p = n\,(m/N)$

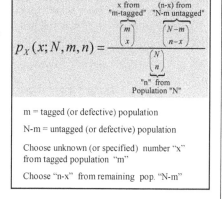

$$p_X(x; N, m, n) = \frac{\binom{m}{x}\binom{N-m}{n-x}}{\binom{N}{n}}$$

x from "m-tagged" $\quad$ (n-x) from "N-m untagged"

"n" from Population "N"

m = tagged (or defective) population

N-m = untagged (or defective) population

Choose unknown (or specified) number "x" from tagged population "m"

Choose "n-x" from remaining pop. "N-m"

**The Hypergeometric PMF answers the question "how many successes "x" are drawn from a sub-population of "m" (defectives), when a total of n samples are taken from the general population N?"** This application is also straightforward if we replace the word "defective" by "tagged," but its use to estimate the size of an unknown animal population is somewhat unique. This is done in two steps by first capturing m animals, tagging them, and then placing them back into the general population, thus establishing a fixed number of "defectives" in an unknown population N. After waiting an amount of time sufficient to mix the tagged animals with the general population, a new sample of size n is captured and the number x of tagged animals is counted. If a series of such measurements are taken, their statistics will follow a hypergeometric distribution, $p_X(x;N, m, n) = [^mC_x * {}^{N-m}C_{n-x}] / {}^NC_n$. Equating the sample mean of this data set $\langle x \rangle_{av}$ to the theoretical mean $\mu_X$ leads to $\langle x \rangle_{av} = \mu_X = n(m/N)$, which gives an estimate $N \approx n \cdot m/\langle x \rangle_{av}$ for the size of the unknown population. Alternately, given just a single measurement of x, the known sample size n, and number of tagged animals m, the hypergeometric probability $p_X(x;N, m, n)$ is only a function of N; thus we seek to maximize $p_X(x; N,m,n)$ as a function of N (most probable or maximum likelihood solution). If N were a continuous variable, setting the derivative equal to zero locates the maximum; however, N is discrete, so we instead consider the ratio of successive terms p(x,N)/p(x,N-1). If there is to be a maximum, then the function $p_X(x; N,m,n)$ must first increase with N which means ratio>1, it then reaches a maximum (ratio=1), and finally decreases which means ratio<1. Hence, the transition point with ratio=1, defines the maximum and hence we set (N-m)(N-n) = N(N-m-n+x), or upon expanding $N^2-(m+n)N +mn = N^2 -N(n+m-x)$, we have mn = Nx. This can be re-written in the suggestive form m/N = x/n which has the simple interpretation that the fraction of tagged animals in the general population m/N is approximately equal to that ratio x/n in the captured sample population. Thus an estimate of the population is N = mn/x which is equivalent to our previous result if we let $x \rightarrow \langle x \rangle_{av}$.

## 7.2.7 Poisson RV Examples

# Poisson RV Examples

**# "x" Succ. in N→∞ Trials**

### Poisson PMF

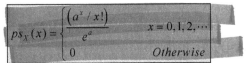

$$ps_X(x) = \begin{cases} \dfrac{(a^x / x!)}{e^a} & x = 0, 1, 2, \cdots \\ 0 & Otherwise \end{cases}$$

Poisson "parameter" $a = \lambda t = rate * time$

Poisson rate $\lambda = number / "time"$ $\begin{cases} errors / page \\ defects / area, etc. \end{cases}$

### Poisson Properties

$$a = \lim_{\substack{n \to \infty \\ p \to 0}}(n \cdot p) = \text{(aver. arrival rate)*time} = \lambda \cdot t$$

$$E[X] = \lim_{\substack{n \to \infty \\ p \to 0}}(n \cdot p) = a$$

$$var(X) = \lim_{\substack{n \to \infty \\ p \to 0}}(n \cdot p \cdot q) = \lim_{\substack{n \to \infty \\ p \to 0}}(\underbrace{n \cdot p}_{\to a} \cdot \underbrace{(1 - p)}_{\to 1}) = a$$

---

### Alpha Particle Decay

**Given Decay rate:** $\lambda = 3.2/sec/gr$
**Find prob. "no more than"** 2 particles decay from in $1gr$ of material in $1sec$

Poisson rate: $\lambda = 3.2/sec/gr$
"Time" interval: $t = 1$ sec-gr

$a = \lambda t = 3.2 /sec/gr * 1 gr * 1 sec = 3.2$

$$P\{X \le 2\} = ps(X = 0) + ps(X = 1) + ps(X = 2)$$
$$= \frac{((3.2)^0 / 0!)}{e^{3.2}} + \frac{((3.2)^1 / 1!)}{e^{3.2}} + \frac{((3.2)^2 / 2!)}{e^{3.2}} = .3799$$

### Earthquake Prediction

**On average 2 Earthquakes per week**
**Find probability "at least"** 3 in next 2 wks

Poisson rate: $\lambda = 2/wk$
Time interval: $t = 2$ wks

$a = \lambda t = (2/wk) * 2 wk = 4$

$$P\{X > 3\} = 1 - P\{X \le 2\}$$
$$= 1 - \left[ \frac{((4)^0 / 0!)}{e^4} + \frac{((4)^1 / 1!)}{e^4} + \frac{((4)^2 / 2!)}{e^4} \right] = .762$$

---

The Poisson PMF gives the probability for exactly x-successes as $(a^x/x!)/e^a$; it is a discrete distribution with an infinite number of terms. The sum of the PMF over x= 0, 1, 2,...,∞ is just a sum of all terms in the Taylor expansion for $e^a$ divided by $e^a$ and hence yields unity. Note also that that the Poisson PMF has its mean and variance both equal to *a* so that *the larger the mean, the larger the variance.* The Poisson PMF is the limiting form of a binomial PMF for a large number of trials with a small probability of success; in fact the Poisson parameter *a* is Lim $_{p \to 0, n \to \infty}$ $(np)$ = a ≡ λ*t and we readily identify a = λt with *np*, the average number of successes for *n* Bernoulli trials. The Poisson PMF models many physical processes such as nuclear decay, earthquake prediction, typographical errors, *etc.*; the rate λ characterizes the process and the time interval *t* is the length of the data run. Thus, for example, if the rate λ represents ".03 typos/page" and we analyze a "100 page" run, then *a* = 3; doubling the run to 200 pages yields a *different* PMF with *a* = 6. The two examples on the slide are summarized below:

**Alpha Particle Decay:** the average #of alpha particles that decay per sec per gram is λ =3.2/sec/gr and this gives a Poisson parameter a= (3.2/sec/gr)*1sec*1gr =3.2. Thus, given 1 gram of material, the probability that no more than 2 alpha particles appear in a 1 sec interval is the sum of the 1st three Taylor series terms of exp(3.2) divided by exp(3.2) which gives .3799 as shown on the slide.

**Earthquake Prediction:** λ =2/wk, so for a 2 week run, *a* = (2/wk)*2 wks = 4 and the probability of at least 3 earthquakes in *two weeks* time is found to be 76.2% by subtracting the sum of four terms from unity. Note that doubling the run to *four weeks* gives a = 8 and the probability increases to 98.6%.

## 7.3 PMFs for Sums of Independent RVs and Convolution

---

# PMFs for Sums of *Independent* RVs and Convolution

Sum Variable : $Z = X + Y$;

Given $p_X(x) \, \& \, p_Y(y)$     Find $p_Z(z)$

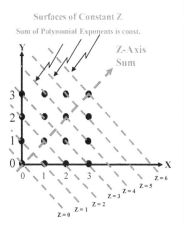

Surfaces of Constant Z
Sum of Polynomial Exponents is const.

$p_{X,Y}(x,y) = p_X(x) \cdot p_Y(y)$ "factors" (*Independence*) *for* all $x, y$

$$p_Z(z) = \sum_{all(x,y):\, x+y=z} \sum p_{X,Y}(x,y) = \sum_{all\, k} p_{X,Y}(x=k, y=z-k)$$

$$p_Z(z) = \sum_k p_X(x=k) \cdot \underset{\text{Independence}}{\underbrace{p_Y(y=z-k)}} = \underset{\text{convolution}}{\underbrace{(p_X * p_Y)[z]}}$$

$$p_Z(z) = \underset{\equiv(p_X * p_Y)[z]}{\underbrace{\sum_k p_X(k) \cdot p_Y(z-k)}} \underset{k \to z-k}{=} \underset{\equiv(p_Y * p_X)[z]}{\underbrace{\sum_k p_X(z-k) \cdot p_Y(k)}}$$

---

Equivalent to Polynomial Multiplication with PMFs = Polynomial Coefficients

$$Poly_X(\lambda) \cdot Poly_Y(\lambda) = Poly_Z(\lambda)$$

$$\sum_{j=0}^{3} p_X(j)\lambda^j \sum_{k=0}^{3} p_Y(k)\lambda^k = \sum_{j=0}^{3}\sum_{k=0}^{3} p_X(j) \cdot p_Y(k)\lambda^{j+k}$$

$$\underset{k=l-j}{=} \sum_{j=0}^{3} \sum_{l=j}^{j+3} p_X(j)p_Y(l-j)\lambda^l = \sum_{l=0}^{6} \underset{=p_Z(l)\ (\text{convolutional sum})}{\underbrace{\left( \sum_{j=0}^{3} p_X(j)p_Y(l-j) \right)}} \lambda^l \quad \text{with} \quad p_Y(l-j<0)=0$$

---

Two independent random variables X and Y are completely characterized by their individual PMFs $p_X(x)$ and $p_Y(y)$. Now, if we take the sum $Z = X+Y$ the question is *"how do we find the PMF $p_Z(z)$ that characterizes the sum variable Z?"* By definition, $p_Z(z)$ is the sum of $p_{XY}(x,y)$ over all $(x,y)$ values satisfying the constraint $x + y = z = $ constant; solving this constraint for y and substituting back into the joint PMF yields $p_{XY}(x, z-x)$, which then factors into the product $p_X(x) \cdot p_Y(z-x)$ because of independence. Upon summing this product over all values of x we are left with a function of z alone. A sum of this form is called a convolution, and is equivalent to the multiplication of two polynomials whose coefficients are the values of the individual PMFs. Taking a PMF with just 4 values we have two degree n=3 polynomials as follows:

$$Poly_X = p_X(0)\lambda^0 + p_X(1)\lambda^1 + p_X(2)\lambda^2 + p_X(3)\lambda^3$$
$$Poly_Y = p_Y(0)\lambda^0 + p_Y(1)\lambda^1 + p_Y(2)\lambda^2 + p_Y(3)\lambda^3$$

Multiplying and collecting terms with the same exponent yields a new polynomial $Poly_Z$ of degree 6 whose coefficients are the values for the sum variable PMF, $p_Z(z=\ell)$. The surfaces of constant Z are the sum lines similar to those for a pair of 4-sided dice except the numbers on the faces are {0,1,2,3} instead of {1,2,3,4}.

The X-Y coordinate system graphic shows the formation of the $6^{th}$ degree product polynomial $Poly_Z$ as a sum of products of the $Poly_X$ and $Poly_Y$ coefficients along the dashed-lines at $-45°$. These are the lines of constant sum labeled z ={0,1,2,3,4,5,6} in the figure. The direct algebraic multiplication of the two polynomials and separation by order of the product polynomial is given by the formal result on the bottom of the slide.

## 7.3.1 Polynomial Multiplication is Convolution

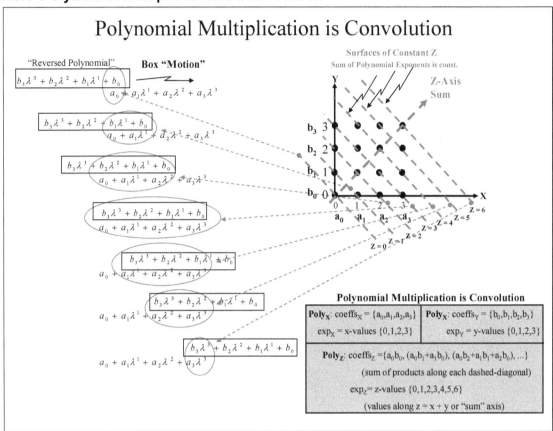

This slide illustrates a geometric interpretation of convolution by taking a descending order polynomial and sliding it over a stationary one to produce a convolutional sum term-by-term as the product exponent *increments by one* at each step of the motion to the right.

This may also be visualized by actually placing the polynomial coefficients along the X and Y coordinate axes; that is, we place the $\text{Poly}_X$ coefficients $\{a_0, a_1, a_2, a_3\}$ along x and the $\text{Poly}_Y$ coefficients $\{b_0, b_1, b_2, b_3\}$ along y as shown. These sets of coefficients are multiplied "as if they were polynomials along their respective axes" and terms with the same exponent are collected along the "constant sum" surfaces Z=0,1,2,3,4,5,6 . This process is equivalent to expanding the polynomial product:

$$(a_0\lambda^0 + a_1\lambda^1 + a_2\lambda^2 + a_3\lambda^3 ) * (b_0\lambda^0 + b_1\lambda^1 + b_2\lambda^2 + b_3\lambda^3 )$$

The motion of the upper polynomial across the stationary one picks off the diagonal sum contributions as indicated by the dashed arrows pointing from the different stages of the box motion to the coordinate diagram. Thus, for example, Z=3 has four contributions $(a_0b_3 + a_1 b_2 + a_2b_1 + a_3b_0)\,\lambda^3$ which sums products of the coefficients corresponding to the 4 points along the diagonal labeled Z=3.

Note that this procedure can also be viewed as a transformation from (X, Y) to sum-difference (Z, D) coordinates; in this case, the polynomial multiplication process collapses the joint $p_{ZD}(z, d)$ onto the sum z-axis yielding the marginal distribution $p_Z(z)$.

## 7.3.1.1 PMF of Dice Sum $S=D_1+D_2$ using Convolution

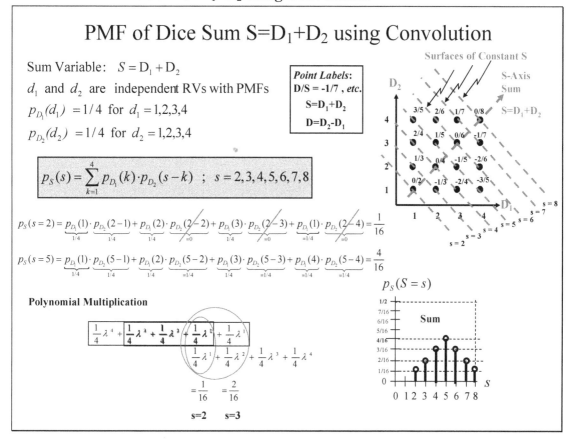

The convolution formula can be applied to a pair of 4-sided dice to find the PMF $p_S(s)$ for the sum random variable $S=D_1+D_2$. We note that each die can only take on values $d_1$, $d_2$ = {1, 2, 3, 4} while the sum S takes on values s= 2,3,4,5,6,7,8; thus care must be taken when evaluating the formula, because of the index "s-k" takes on values for which $p_{D2}(s-k)$ is zero. Two sample calculations are exhibited and we see that for s=2 only one of the 4 terms in the expansion is non-zero, while for s=5 all four terms are non-zero.

Alternately, from the labeled coordinate representation of the joint distribution $p_{D_1D_2}(d_1,d_2)$ shown in the slide, the values may be easily picked off and summed (collapsed) onto the sum axis (Z-axis) to directly yield the PMF $p_S(s)$ for s={2,3,4,5,6,7,8}. The labeling of points is D/S where D is the difference coordinate $D=D_2-D_1$ and S is the sum coordinate S= $D_1+D_2$; thus, the point -2/4 corresponds to D= -2 and S=4.

Finally, the polynomial multiplication performed by sliding the "box" containing the descending order polynomial against the stationary ascending polynomial creates the convolution term-by-term by summing products as the upper and lower terms register with each other. The resulting sum PMF $p_S(s)$ is shown in the lower right stick plot graph.

## 7.3.2 Sum of Two Binomials is Binomial by Convolution

# Sum of Two Binomials is Binomial by Convolution

**Sum of Two Binomial Distributed RVs is also Binomial**

$X$ : is a binomial $b_X(x; n, p)$ : $\quad p_X(x) = \binom{n}{x} p^x q^{n-x}$

$Y$ : is a binomial $b_Y(y; m, p)$ : $\quad p_Y(y) = \binom{m}{y} p^y q^{m-y}$

$\Rightarrow$

$Z = X + Y$ : is a binomial $b_Z(z; (n+m), p)$ :

$$p_Z(z) = \binom{(n+m)}{z} p^z q^{(n+m)-z}$$

**Proof:** $\quad p_Z(z) = \displaystyle\sum_{k=0}^{m+n} b_X(X = k; n, p) \cdot b_Y(Y = z - k; m, p)$

$$= \sum_{k=0}^{m+n} \binom{n}{k} p^k q^{n-k} \cdot \binom{m}{z-k} p^{z-k} q^{m-(z-k)} = \left\{ \sum_{k=0}^{m+n} \binom{n}{k}\binom{m}{z-k} \right\} \cdot p^z q^{(m+n)-z} = \binom{(n+m)}{z} \cdot p^z q^{(n+m)-z}$$

*Hypergeometric Identity (6) Slide#2-22*

**n Bernoulli trials yields a Binomial**

**Proof by Induction:** *Bernoulli RV is a Binomial RV with one draw, i.e., n=1:*

Apply above result for two binomials with n=1, m=1

$X_1$: $b_{X_1}(x_1; 1, p)$ : $\quad p_{X_1}(x_1) = \binom{1}{x_1} p^{x_1} q^{1-x_1}$

$X_2$: $b_{X_2}(x_2; 1, p)$ : $\quad p_{X_2}(x_2) = \binom{1}{x_2} p^{x_2} q^{1-x_2}$

$\Rightarrow$

$Z_1 = X_1 + X_2$ : is a binomial
$b_{Z_1}(z_1; 1+1, p) = b_{Z_1}(z_1; 2, p)$

$$p_{Z_1}(z_1) = \binom{2}{z_1} p^{z_1} q^{2-z_1}$$

$Z_2 = Z_1 + X_3$ : is a binomial
$b_{Z_2}(z_2; 2+1, p) = b_{Z_2}(z_2; 3, p)$

$$p_{Z_2}(z_2) = \binom{3}{z_2} p^{z_2} q^{3-z_2}$$

Add 3$^{rd}$ Bernoulli $X_3$ to above result to find

Induction: Assume true for n Bernoulli RVs and show true for n+1 Bernoulli RVs    Q.E.D

We give a formal proof that the sum of two independent binomials $b_Z(x_1; n,p)$ and $b_Z(x_2; m,p)$ with different orders "n" and "m" and with the same single trial probability of success "p" is again a binomial $b_Z(z; m+n, p)$ for the sum random variable $Z=X_1+X_2$ with order m+n. The proof shown in the upper panel proceeds by taking the convolutional sum of the two binomials $b_{X_1}(x_1;n,p)$ and $b_{X_2}(x_2;m,p)$ which yields a product of the two combinatorial coefficients and a sum of their exponents. Using the fundamental Hypergeometric identity (6) (Slide#2-22) collapses the sum $\sum_k {}^nC_k \, {}^mC_{z-k} = {}^{m+n}C_z$ to yield the binomial $b_Z(z; m+n, p)$ .

In the lower panel, we apply the above result that the sum of two binomials is again a binomial to prove that the sum of n Bernoulli trials is a Binomial of degree n by induction. We first recognize that a binomial RV with n=1 is a Bernoulli RV and directly apply the above result to two Bernoulli RVs $X_1$, $X_2$ to obtain a binomial RV, $b_{Z_1}(z_1;2,p)$ with m+n=1+1=2. Adding a 3$^{rd}$ Bernoulli RV to the resulting binomial RV $b_{Z_1}(z_1;2,p)$ yields a binomial RV $b_{Z_2}(z_2;3,p)$ of order 3. Continuing in this fashion, we can arrive at a result for n =10 which states the sum of 10 Bernoulli RVs is a binomial of order 10. Formal proof by induction proceeds by assuming that this result is true for arbitrary "n" and then showing that it is true for the next order "n+1".

## 7.3.3 Moment Generating Function / Transform Technique

# Moment Generating Function / Transform Technique

**Moment Generating Function:** $\qquad \varphi_X(t) \equiv E[e^{t \cdot X}] = \sum_x e^{t \cdot X} p_X(x)$ $\qquad$ *Take derivatives wrt "t" and evaluate at t=0:*

1st Moment ( Mean ) $\quad \varphi_X'(t)\big|_{t=0} = \dfrac{d}{dt} E[e^{t \cdot X}]\big|_{t=0} = E[X \cdot e^{t \cdot X}]\big|_{t=0} = E[X] = \mu_X$ $\qquad$ nth Moment

2nd Moment $\qquad \varphi_X''(t)\big|_{t=0} = \dfrac{d^2}{dt^2} E[e^{t \cdot X}]\big|_{t=0} = \dfrac{d}{dt} E[X \cdot e^{t \cdot X}]\big|_{t=0} = E[X^2 \cdot e^{t \cdot X}]\big|_{t=0} = E[X^2]$ $\qquad E[X^n] = \dfrac{d^n}{dt^n} E[e^{t \cdot X}]\big|_{t=0}$

---

**Convolution of PMFs yields Product of Moment Generating Functions:**

Given two Indep. RVs: X,Y $\qquad$ **Proof:** $\quad \varphi_Z(t) = E[e^{t(X+Y)}] = \sum_x \sum_y \underbrace{p_{XY}(x,y)}_{=p_X(x)p_Y(y)} \cdot e^{t(X+Y)}$

Form their sum: Z = X + Y

Transform of their sum is $\qquad\qquad\qquad = \sum_x p_X(x) e^{tX} \sum_y p_Y(y) e^{tY} = \varphi_X(t) \cdot \varphi_Y(t)$

$\varphi_Z(t) = \varphi_X(t) \cdot \varphi_Y(t)$ $\qquad\qquad\qquad\qquad\qquad \underbrace{\phantom{\sum_x p_X(x) e^{tX}}}_{=\varphi_X(t)} \underbrace{\phantom{\sum_y p_Y(y) e^{tY}}}_{=\varphi_Y(t)}$

---

| **Equivalent Representations of RV Properties** | **Example: Sum of Two Binomials** $b_X(x;n,p)$ & $b_Y(y;m,p)$ |
|---|---|
| **RVs: X, Y** $\qquad$ **X + Y = Z** $\qquad$ **Sum RV: Z** | *Moment Generating Function for each Binomial* |
| *Transform Each* $\qquad\qquad$ *Product of Transforms* | $\varphi_X(t) = \sum_x e^{tx} \binom{n}{x} p^x q^{n-x} = \sum_x \binom{n}{x}(pe^t)^x q^{n-x} = (pe^t + q)^n$ |
| $\phi_X(t) \Leftrightarrow p_X(x)$ $\qquad\qquad \varphi_Z(t) = \varphi_X(t) \cdot \varphi_Y(t)$ | $\varphi_Z(t) = \varphi_X(t) \cdot \varphi_Y(t) = (pe^t + q)^n \cdot (pe^t + q)^m = \underbrace{(pe^t + q)^{m+n}}_{\text{binomial "n+m"}}$ |
| $\varphi_Y(t) \Leftrightarrow p_Y(y)$ | |
| *Transform Back* | |
| $\varphi_Z(t) \Leftrightarrow p_Z(z)$ | *Transform Back* $\qquad \varphi_Z(t) \Rightarrow p_Z(z) \Rightarrow b(z;(n+m),p)$ |

The moment generating function for random variable X is defined as the expectation value $\phi_X(t)=E[e^{Xt}]$ and has the property that its derivative evaluated at t=0 is the mean $\mu_X$, further differentiation yields higher order moments of the distribution. Thus the moment generating function constitutes an alternate characterization of a RV. The center panel of this slide proves that the generating function for the sum variable Z=X+Y is simply the product of the two generators, *viz.*, $\phi_Z(t) = \phi_X(t) \phi_Y(t)$. This allows us to replace the tedious convolution calculation with a simple product. However, there is no general procedure to make the inverse transform, so the resulting product generator needs to be recognized as belonging to a known PMF for it to be useful.

As an example, we can easily verify that the moment generating function for a Bernoulli RV is

$$\phi_X(t) = \sum_{X=0}^{1} e^{X \cdot t} p_X(x) = e^{0 \cdot t} q + e^{1 \cdot t} p = q + pe^t \qquad \text{Bernoulli}$$

Since a Binomial RV results from summing "n" Bernoulli RVs, its moment generating function is simply the product of n Bernoulli generators. Thus for the sum variable $Z=X_1 + X_2 + ... + X_n$, we have the generating function $\phi_Z(t) = (\phi_X(t))^n = (q + p e^t)^n$ and the corresponding PMF is of course a binomial $b_X(x;n,p)$. Note that the moment generating function for a binomial PMF can also be derived directly as follows:

$$\phi_X(t) = \sum_{X=0}^{n} e^{X \cdot t} p_X(x) = \sum_{x=0}^{n} e^{x \cdot t} \,^n C_k p^x q^{n-x} = \sum_{x=0}^{n} \,^n C_k (pe^t)^x q^{n-x}$$

Upon substituting a=q and b=pe$^t$ in the last sum, it is easily recognized to be the binomial expansion of $(a+b)^n$, so we have

$$\phi_X(t) = (q + pe^t)^n \qquad \text{Binomial}$$

Using the generating functions $\phi_X(t)=(q+pe^t)^n$ and $\phi_Y(t)=(q+pe^t)^m$ for binomials X: $b_X(x;n,p)$ and Y: $b_Y(y;m,p)$, respectively, their product yields $\phi_Z(t)= \phi_X(t) \phi_Y(t) = (q+pe^t)^n \cdot (q+pe^t)^m =(q+pe^t)^{n+m}$, which is immediately recognized as the generator for a binomial Z: $b_Z(z; \mathbf{m+n},p)$. Note how simple this is compared to the direct proof (Slide#7-20) which required Hypergeometric (binomial) identity (6) to evaluate the expectation sum.

## 7.3.3.1 Geometric Mean and Variance from Generating Function

# Geometric Mean and Variance from Generating Function

| Geometric PMF | $p_X(x) = pq^{x-1} \quad x = 1, 2, \cdots$ | Generating Function | $\varphi_X^{Geom}(t) = \dfrac{pe^t}{(1 - qe^t)}$ |
|---|---|---|---|

*Derivation of Generating Function*

$$\varphi_X(t) = \sum_x e^{tx} pq^{x-1} = pq^{-1} \sum_{x=1}^{\infty} \left(qe^t\right)^x = \frac{p}{q} qe^t \left[1 + qe^t + \left(qe^t\right)^2 + \cdots\right] = \frac{pe^t}{1 - qe^t}$$

*Calculation of Moments*

$$E[X^1] = \frac{d}{dt}\varphi_X(t)\Big|_{t=0} = \frac{pe^t(1 - qe^t) - pe^t(-qe^t)}{(1 - qe^t)^2} = \frac{pe^t}{(1 - qe^t)^2}\Big|_{t=0} = \frac{p}{(1 - q)^2} = \boxed{\frac{1}{p}} \quad \textbf{Mean}$$

$$E[X^2] = \frac{d^2}{dt^2} E[e^{t \cdot X}]\Big|_{t=0} = \frac{d}{dt}\left\{\frac{pe^t}{(1 - qe^t)^2}\right\}\Big|_{t=0} = \frac{pe^t(1 - qe^t)^2 - pe^t 2(-qe^t)(-q)}{(1 - qe^t)^4}$$

$$= \frac{p(1 - q)^2 - 2qp(1 - q)}{(1 - qe^t)^4} = \frac{1 + q}{p^2}$$

$$\mathrm{var}(X) = E[X^2] - E[X]^2 = \frac{1 + q}{p^2} - \left(\frac{1}{p}\right)^2 = \boxed{\frac{q}{p^2}} \quad \textbf{Variance}$$

In this slide we develop the moment generating function for the geometric PMF and then compute its mean and variance by simply computing the first two moments using the derivatives of the generating function evaluated at t = 0. The definition of $\phi_X(t)$ as the expectation value of the exponential $e^{Xt}$ using the geometric PMF requires us to sum the product $e^{xt} * pq^{x-1}$ over all $x = 1, 2, 3, \cdots, \infty$. A constant term $p*q^{-1}$, is factored outside, leaving a sum of terms $(qe^t)^x$; these terms are written out on the 2nd line of the calculation and upon taking out the factor $qe^t$ there remains an infinite sum corresponding to a geometric sequence $1 + s + s^2 + s^3 + \ldots = 1/(1-s)$, where $s = qe^t$. The expression for $\phi_X(t)$ now takes the form $p*q^{-1} * qe^t/(1 - qe^t)$ which reduces to $\phi_X(t) = pe^t/(1 - qe^t)$ as shown on the slide.

Evaluation of the 1st and 2nd moments involves straightforward differentiation of $\phi_X(t)$ with respect to the variable "t" and subsequent evaluation at t=0. Note the obvious fact that the next derivative must be taken before evaluation at t=0; otherwise there will be no variable "t" to differentiate a second time. Also note that in order to calculate the var(X) we need both the 1st and 2nd moments of the distribution.

## 7.3.3.2 Negative Binomial Mean and Variance from Generating Function

### Negative Binomial Mean and Variance from Generating Fcn

| Negative Binomial PMF | $p_X(x) = \binom{x-1}{r-1} p^{r-1} q^{x-r} \cdot \underbrace{p}_{\substack{\text{succ. on} \\ \text{next trial}}}$ <br> $\underbrace{\phantom{p_X(x)}}_{(r-1)\text{succ. in } (x-1) \text{ trials}}$ | Generating Function | $\varphi_X^{NegBinom}(t) = \left(\dfrac{pe^t}{(1-qe^t)}\right)^r = \left(\varphi_X^{Geom}(t)\right)^r$ |
|---|---|---|---|

**Derivation of Generating Function**

$$\varphi_X(t) = \sum_{x=r}^{\infty} e^{tx} \binom{x-1}{r-1} p^r q^{x-r} = e^{+tr} \sum_{x=r}^{\infty} e^{tx} \binom{x-1}{r-1} p^r \left(qe^t\right)^{x-r}$$

$$= e^{+tr} \left\{ \binom{r-1}{r-1} p^r \left(qe^t\right)^{r-r} + \binom{r}{r-1} p^r \left(qe^t\right)^1 + \binom{r+1}{r-1} p^r \left(qe^t\right)^2 + \cdots \right\}$$

$$= p^r e^{+tr} \cdot \underbrace{\left\{ 1 + r\left(qe^t\right)^1 + \frac{r(r+1)}{2}\left(qe^t\right)^2 + \cdots \right\}}_{= 1/(1-qe^t)^r} = \left(\frac{pe^t}{1-qe^t}\right)^r$$

**Calculation of 1st Moment (Mean)**

$$E[X] = \frac{d}{dt} \varphi_X(t) \bigg|_{t=0} = \frac{d}{dt}\left(\left(\frac{pe^t}{1-qe^t}\right)^r\right)\bigg|_{t=0} = r\left(\frac{pe^t}{1-qe^t}\right)^{r-1} \frac{pe^t(1-qe^t) - pe^t(-qe^t)}{(1-qe^t)^2}\bigg|_{t=0}$$

$$= r \frac{\left(pe^t\right)^r}{\left(1-qe^t\right)^{r+1}}\bigg|_{t=0} = \frac{rp^r}{(1-q)^{r+1}} = \frac{r}{p} = \boxed{r \cdot \left(\frac{1}{p}\right)} \qquad \text{r times the geometric mean!}$$

| Negative Binomial Expansion | $(1-qe^t)^{-r} = 1^{-r} + (-r)1^{-r-1}(-qe^t)^1 + \dfrac{(-r)(-r-1)}{2}1^{-r-2}(-qe^t)^2 + \cdots$ |
|---|---|
| | $= 1 + rqe^t + \dfrac{(r)(r+1)}{2}(qe^t)^2 + \dfrac{(r)(r+1)(r+2)}{3!}(qe^t)^3 + \cdots$ |

In this slide we develop the generating function for the Negative binomial PMF and then compute its mean using the 1st derivative of the generating function evaluated at t=0. The calculation of the 2nd moment and variance are left as an exercise. The definition of $\phi_X(t)$ as the expectation value of the exponential $e^{Xt}$ using the negative binomial PMF requires us to sum the product $e^{xt} * {}^{x-1}C_{r-1} p^r q^{x-r}$ over all x. Note that for the negative binomial, we do not start the sum at x=1, but rather at x=r (which is the minimum number of trials for r-successes); thus x ranges over the values x = r, r+1, r+2, ⋯,∞. The factor $e^{tx}$ inside the sum is re-written in the form $e^{tr}*e^{t(x-r)}$ so that it can be conveniently combined with the term $q^{x-r}$ to yield $e^{tr}* (qe^t)^{(x-r)}$. The first term of the latter expression may be taken outside the sum leaving us with a sum of products of the form ${}^{x-1}C_{r-1} p^r (qe^t)^{(x-r)}$; we explicitly write out a few terms of this sum and display them within the curly brackets {}. At the bottom of the slide an expansion of the negative binomial expression $(1- qe^t)^{-r}$ is shown to generate precisely the {} terms; thus, replacing the infinite sum with $(1- qe^t)^{-r}$ yields the generating function as $\phi_X(t) = [pe^t /(1- qe^t)]^r$.

Taking the 1st derivative d/dt ($\phi_X(t)$) and evaluating it at t=0, gives the mean E[X] = r (1/p) which is "r" times the mean for a geometric RV. This is easily understood because the *negative binomial* RV is just the sum of "r" independent *geometric* RVs so its mean must be r times 1/p. Moreover, this relation also follows from the fact that the generating function for the negative binomial is just the product of r geometric generating functions, so the derivative brings down the exponent r. Not surprisingly, the variance for the negative binomial RV is also "r" times that for a geometric RV which is $r \cdot (q/p^2)$. These two distributions {geometric, negative binomial} are related in the same way as the {Bernoulli, binomial} pair (see slides 7-2, 7-3, 7-4 for comparisons and related trees).

# 8 References

1. "*A First Course in Probability*, 7<sup>th</sup> Ed., " Ross, Sheldon, Prentice-Hall, 2006.

2. "*An Introduction to Applied Probability*," Roberts, Robert A., Addison-Wesley, 1992.

3. "*Probability Models and Applications*, 2$^{nd}$ Ed.," Olkin, I., Gleser, L.J., Derman, C., Prentice-Hall, 1994.

4. "*Probability, Statistics, and Random Processes for Electrical Engineering*, 3$^{rd}$ Ed.," Leon-Garcia, Leon, Prentice-Hall, 2008.

5. "*Introduction to Mathematical Statistics*, 5$^{th}$ Ed.," Hogg, Robert V., Craig, Allen T., Prentice-Hall, 1995.

# Index

## 9 Index

# Index

# Index

# Index

CPSIA information can be obtained
at www.ICGtesting.com
Printed in the USA
FSHW011024250719
60377FS

9 781481 282062